T0255585

Lecture Notes in Computer Science 13872

The series Lecture Notes in Computer Science (LNCS), including its subseries Lecture Notes in Artificial Intelligence (LNAI) and Lecture Notes in Bioinformatics (LNBI), has established itself as a medium for the publication of new developments in computer science and information technology research, teaching, and education.

LNCS enjoys close cooperation with the computer science R & D community, the series counts many renowned academics among its volume editors and paper authors, and collaborates with prestigious societies. Its mission is to serve this international community by providing an invaluable service, mainly focused on the publication of conference and workshop proceedings and postproceedings. LNCS commenced publication in 1973.

Georgiana Caltais · Christian Schilling
Editors

Model Checking Software

29th International Symposium, SPIN 2023
Paris, France, April 26–27, 2023
Proceedings

Editors
Georgiana Caltais
University of Twente
Enschede, The Netherlands

Christian Schilling
Aalborg University
Aalborg, Denmark

ISSN 0302-9743 ISSN 1611-3349 (electronic)
Lecture Notes in Computer Science
ISBN 978-3-031-32156-6 ISBN 978-3-031-32157-3 (eBook)
https://doi.org/10.1007/978-3-031-32157-3

This Springer imprint is published by the registered company Springer Nature Switzerland AG
The registered company address is: Gewerbestrasse 11, 6330 Cham, Switzerland

Preface

This volume contains the proceedings of the 29th International Symposium on Model Checking Software, SPIN 2023, co-located with ETAPS 2023 in Paris, at Sorbonne Université, on April 26–27, 2023. Since 1995, the event has evolved and has been consolidated as a reference symposium in the area of formal methods related to model checking.

The SPIN 2023 edition requested regular and short papers in the following areas: formal verification techniques for automated analysis of software; formal analysis for modeling languages such as UML/state charts; formal specification languages, temporal logic, design-by-contract; model checking; analysis of concurrent software; automated theorem proving, including SAT and SMT; verifying compilers; abstraction and symbolic execution techniques; static analysis and abstract interpretation; combination of verification techniques; modular and compositional verification techniques; verification of timed and probabilistic systems; automated testing using advanced analysis techniques; combination of static and dynamic analyses; program synthesis; derivation of specifications, test cases, or other useful material via formal analysis; case studies of interesting systems or with interesting results; engineering and implementation of software verification and analysis tools; benchmark and comparative studies for formal verification and analysis tools; formal methods of education and training; insightful surveys or historical accounts on topics of relevance to the symposium; relevant tools and algorithms for modern hardware, such as parallel, GPU, TPU, cloud, and quantum; formal techniques to model and analyze societal and legal systems; formal analysis of learned systems.

The symposium attracted 21 submissions, which were each carefully single-blind reviewed by three Program Committee (PC) members. The selection process included further online discussion open to all PC members. As a result, 9 full papers and 2 short papers were selected for presentation at the symposium and publication in Springer's proceedings.

The program included 4 invited talks: Simon Gay, University of Glasgow: Model-Checking Quantum Computing Systems; Joost-Pieter Katoen, RWTH Aachen & University of Twente: The Probabilistic Model Checker Storm; Raúl Pardo, IT University of Copenhagen: Formal Verification of Privacy Policies for Social Networks; Caterina Urban, INRIA: Interpretability-Aware Verification of Machine Learning Software.

We would like to thank all the authors who submitted papers, the Steering Committee, the PC, the additional reviewers, the invited speakers, the participants, and the organizers of the co-hosted events for making SPIN 2023 a successful event.

March 2023

Georgiana Caltais
Christian Schilling

Organization

Program Committee Chairs

Georgiana Caltais	University of Twente, The Netherlands
Christian Schilling	Aalborg University, Denmark

Steering Committee

Dragan Bosnacki	Eindhoven University of Technology, The Netherlands
Susanne Graf	Verimag, France
Gerard Holzmann	Nimble Research, USA
Stefan Leue	University of Konstanz, Germany
Jaco van de Pol	Aarhus University, Denmark
Neha Rungta	Amazon Web Services, USA
Willem Visser	Stellenbosch University, South Africa

Program Committee

Guy Avni	University of Haifa, Israel
Kyungmin Bae	Pohang University of Science and Technology, South Korea
Roberto Bagnara	University of Parma, Italy
Ezio Bartocci	TU Vienna, Austria
Sergiy Bogomolov	Newcastle University, UK
Dragan Bosnacki	Eindhoven University of Technology, The Netherlands
Krishnendu Chatterjee	IST Austria, Austria
Bernd Finkbeiner	CISPA Helmholtz Center for Information Security, Germany
Miriam García Soto	Complutense University of Madrid, Spain
Mirco Giacobbe	University of Birmingham, UK
Arnd Hartmanns	University of Twente, The Netherlands
Klaus Havelund	Jet Propulsion Laboratory, USA
Gerard Holzmann	Nimble Research, USA
Taylor T. Johnson	Vanderbilt University, USA

Bernhard Kragl	Amazon Web Services, Austria
Axel Legay	University of Louvain, Belgium
Stefan Leue	University of Konstanz, Germany
Anna Lukina	TU Delft, The Netherlands
Sergio Mover	École Polytechnique, France
Rajagopal Nagarajan	Middlesex University, UK
Daniel Neider	University of Oldenburg, Germany
Corina Pasareanu	NASA Ames & CMU, USA
Doron Peled	Bar Ilan University, Israel
Andreas Podelski	University of Freiburg, Germany
Christian Schilling	Aalborg University, Denmark
Natasha Sharygina	University of Lugano, Switzerland
Jiří Srba	Aalborg University, Denmark
Michael Tautschnig	Amazon Web Services & Queen Mary University of London, UK
Georg Weissenbacher	TU Vienna, Austria

Additional Reviewers

Martin Blicha
Michele Chiari
Dennis Dams
Ignacio D. Lopez-Miguel
Abdelrahman Hekal
Radu Mateescu
Rodrigo Otoni
Paulius Stankaitis
Ennio Visconti
Stefan Zetzsche

Contents

Binary Decision Diagrams

Explicit-State Model Checking

Binary Decision Diagrams

Efficient Implementation of LIMDDs for Quantum Circuit Simulation

Lieuwe Vinkhuijzen[1(✉)], Thomas Grurl[2,3], Stefan Hillmich[3], Sebastiaan Brand[1], Robert Wille[4,5], and Alfons Laarman[1]

[1] Leiden University, Leiden, The Netherlands
l.t.vinkhuijzen@liacs.leidenuniv.nl
[2] Secure Information Systems, University of Applied Sciences Upper Austria, Wels, Austria
[3] Institute for Integrated Circuits, Johannes Kepler University Linz, Linz, Austria
[4] Chair for Design Automation, Technical University of Munich, Munich, Germany
[5] Software Competence Center Hagenberg Gmbh (SCCH), Hagenberg Im Mühlkreis, Austria

Abstract. Realizing the promised advantage of quantum computers over classical computers requires both physical devices and corresponding methods for the design, verification and analysis of quantum circuits. In this regard, decision diagrams have proven themselves to be an indispensable tool due to their capability to represent both quantum states and unitaries (circuits) compactly. Nonetheless, recent results show that decision diagrams can grow to exponential size even for the ubiquitous stabilizer states, which are generated by Clifford circuits. Since Clifford circuits can be efficiently simulated classically, this is surprising. Moreover, since Clifford circuits play a crucial role in many quantum computing applications, from networking, to error correction, this limitation forms a major obstacle for using decision diagrams for the design, verification and analysis of quantum circuits. The recently proposed *Local Invertible Map Decision Diagram* (LIMDD) solves this problem by combining the strengths of decision diagrams and the stabilizer formalism that enables efficient simulation of Clifford circuits. However, LIMDDs have only been introduced on paper thus far and have not been implemented yet—preventing an investigation of their practical capabilities through experiments. In this work, we present the first implementation of LIMDDs for quantum circuit simulation. A case study confirms the improved performance in both worlds for the Quantum Fourier Transform applied to a stabilizer state. The resulting package is available under a free license at https://github.com/cda-tum/ddsim/tree/limdd.

1 Introduction

Quantum computing is a new and drastically different computing paradigm promising to solve certain problems that are intractable for classical computers. Examples of such problems include unstructured search [1–3], integer factorization [4] and quantum chemistry [5]. This computational power is harnessed by using quantum mechanical effects such as *superposition*, where the system can be

G. Caltais and C. Schilling (Eds.): SPIN 2023, LNCS 13872, pp. 3–21, 2023.
https://doi.org/10.1007/978-3-031-32157-3_1

in a linear combination of multiple states, and *entanglement*, where operations on one part of the system can affect other parts as well. In the near term, quantum computers classified as *Noisy Intermediate-Scale Quantum* (NISQ) devices are expected to both deliver empirical evidence of quantum advantage over classical computers as well as solve practical problems. However, the ability to build large quantum computers is not by itself sufficient if there are no means of harnessing their power: we also need tools for the design of quantum circuits, i.e., for simulation, compilation, and verification. The *classical simulation* of quantum computers in particular has been used in service of the verification of quantum algorithms [6–8], and is a way to quantify "the elusive boundary at which a quantum advantage may materialize" [9].

A major challenge in the classical design of quantum systems is that the memory requirements grow exponentially in the number of qubits. Contrary to the classical world, where representing a system state with m classical bits requires only a linear amount of memory, the state of an n-qubit quantum system is described by a vector of 2^n complex numbers. Current estimates indicate that at least hundreds of qubits are required to perform useful tasks on a quantum computer [10]. However, even current super-computing clusters can only handle systems with between 50 and 60 qubits represented as vectors [11]. Therefore, dedicated data structures and design methods which can tackle the exponential complexity of quantum computing need to be developed.

Given that merely representing a quantum state may require an exponential amount of memory with respect to the number of qubits, it comes as no surprise that conducting quantum circuit simulation, for example, is a hard problem. Even more dauntingly, verification of quantum circuits at its heart considers quantum operations and therefore has $2^n \times 2^n$ complexity when implemented naively. Fortunately, due to the characteristics of quantum computing, quantum circuit simulation can help to verify the equivalence of two circuits to a very high degree of confidence, despite the infinite number of possible input states. More precisely, providing an appropriate quantum state as input to the circuits under consideration and checking equivalence of the resulting states will provide a counterexample for non-equivalent circuits with a high probability [7]. Selecting basis states as input, i.e., states without superposition or entanglement, does not always suffice for this purpose, but random *stabilizer states*, introduced next, have been shown to do the trick [6]. This makes quantum circuit simulation a key component of design automation for quantum computing and hence the subject of the current paper.

Stabilizer states are ubiquitous in quantum computing. They are computed by so-called Clifford circuits, a subset of the universal quantum computing gate set [12,13]. For example, stabilizer states include the Bell state and GHZ state. Further, Clifford circuits play an essential role in error correction [14,15], entanglement distillation [16] and are used in one-way quantum computing [17]. Any n-qubit stabilizer state can be represented using memory in the order of $\mathcal{O}(n^2)$ and the non-universal fragment of Clifford circuits can be simulated efficiently by manipulating this representation [12,13]. In fact, stabilizer states capture the essential symmetries of all universal quantum computing states, which is why they also play a key role in the reduction from verification to simulation explained above. For these reasons, it can be argued that any practically efficient classical simulation of (universal) quantum computing should also support Clifford circuits and the stabilizer states they generate. The current work removes this limitation from existing (universal) simulation approaches based on decision diagrams.

Decision Diagrams (DDs) are a tried-and-tested data structure in the world of classical design automation [18–22]. They have also shown promising results in quantum design automation [23–28]. Decision diagrams exploit redundancies in the state vector and operations matrix to enable a compact representation in many cases. Unfortunately, a state-of-the art quantum simulation method called Quantum Multi-valued Decision Diagram (QMDD) [29] does not efficiently represent stabilizer states [27], which poses a serious bottleneck to their adoption, as explained above, but also observed in practice [6]. The recently proposed *Local Invertible Map Decision Diagram* (LIMDD, [27]) addresses this shortcoming. LIMDDs efficiently represent stabilizer states, they simulate Clifford circuits in polynomial time and can efficiently apply many Clifford gates also to non-stabilizer states. However, LIMDDs lack an implementation to demonstrate that their asymptotic advantage also translates to practical use cases.

In this paper, we present an implementation of LIMDDs for universal simulation of quantum circuits (and thus design automation) based on the QMDD package [26,28]. We adapt techniques that are tried and tested in the implementations of both classical and quantum decision diagram packages, and enrich them with special considerations to efficiently handle *Local Invertible Maps* (LIMs). For the first time, this leads to an implementation that realizes LIMDDs, and also demonstrates the potential of LIMDDs. In particular, we show their use for verification through circuit equivalence checking for a case study on the Quantum Fourier Transform (QFT, [30,31]). The results confirm that the more complex LIMDD-based simulator surpasses a state-of-the-art decision-diagram-based simulator for larger instances. The resulting implementation is available at https://github.com/cda-tum/ddsim/tree/limdd under the MIT license.

The remainder of this paper is structured as follows. Section 2 briefly reviews the necessary background on quantum computing and classical quantum circuit simulation. In Sect. 3, we briefly review existing decision diagrams for quantum computing and motivate the need for an efficient LIMDD implementation. Section 4 details the techniques used to enable efficient construction and manipulation of LIMDDs. In Sect. 5, we provide an experimental evaluation showcasing the performance of the proposed implementation. Finally, Sect. 6 concludes the paper.

2 Background

To keep this work self-contained, this section provides the necessary background on quantum computing as well as classical quantum circuit simulation.

2.1 Quantum States and Operations

The basic unit of information in quantum computing is the *quantum bit* or *qubit* [30]. A single-qubit quantum state $|\psi\rangle$ can be described by its amplitude vector $|\psi\rangle = \alpha_0 \cdot |0\rangle + \alpha_1 \cdot |1\rangle$ with complex amplitudes $\alpha_0, \alpha_1 \in \mathbb{C}$. Here, $|0\rangle = [\begin{smallmatrix}1\\0\end{smallmatrix}], |1\rangle = [\begin{smallmatrix}0\\1\end{smallmatrix}]$ denote the two basis states and are analogous to the basis states 0 and 1 in classical computing, in the sense that a classical register containing one bit can be either in state 0 or 1. An amplitude vector must meet the normalization constraint $|\alpha_0|^2 + |\alpha_1|^2 = 1$. If both amplitudes α_0 and α_1 are non-zero, the state is in *superposition.* For a state $|\phi\rangle$, we write $\langle\phi| = |\phi\rangle^\dagger$ to denote its conjugate transpose. Two quantum states $|\phi\rangle, |\psi\rangle$ are considered equal if the absolute value of their in-product equals one, i.e., $|\langle\phi| \cdot |\psi\rangle| = 1$, also written as $|\langle\phi|\psi\rangle| = 1$. We say that $|\phi\rangle, |\psi\rangle$ are approximately the same state if $|\langle\phi|\psi\rangle| \approx 1$. When measuring a state, the probability that a given basis state is the outcome is the squared magnitude of the amplitude of that basis state, i.e., for the state $\alpha_0 \cdot |0\rangle + \alpha_1 \cdot |1\rangle$, the probability of measuring the zero state is $|\alpha_0|^2$. Therefore, in a physical quantum computer, the individual amplitudes are fundamentally non-observable and information about the quantum state can only be extracted through destructive measurement, i.e., after measurement, superposition is destroyed.

A quantum register may consist of multiple qubits. A register $|\phi\rangle$ consisting of n qubits has 2^n basis states $|i\rangle$ with $i \in \{0,1\}^n$, each with a corresponding amplitude α_i, i.e., $|\phi\rangle = \sum_i \alpha_i |i\rangle$. Here a basis state $|i\rangle$ with i a bit string $b_1 b_2 \ldots b_n$ is formed from the single qubit basis vectors above using tensor products $|b_1\rangle \otimes |b_2\rangle \otimes \cdots \otimes |b_n\rangle$, written shortly as $|b_1\rangle |b_2\rangle \ldots |b_n\rangle = |b_1 b_2 \ldots b_n\rangle = |i\rangle$. Here the *tensor product* of a $k \times \ell$ matrix $A = (a_{ij})_{ij}$ and an $n \times m$ matrices $B = (b_{ij})_{ij}$ is the $kn \times \ell m$ matrix $C = (a_{ij} b_{xj})_{ijxy}$. Alternatively, $|\phi\rangle$ can be understood as the pseudo-Boolean function $f \colon \{0,1\}^n \to \mathbb{C}$, defined as $f(i) = \langle i|\phi\rangle = \alpha_i$. The normalization constraint is generalized to $\sum_{0 \le i < 2^n} |\alpha_i|^2 = 1$. A quantum state $|\phi\rangle$ is *entangled* if it cannot be written as a tensor product of single qubit states, i.e., $|\phi\rangle = |\phi_1\rangle \otimes \cdots \otimes |\phi_n\rangle$.

Example 1. Consider the quantum state $1/\sqrt{2} \cdot (|00\rangle + |11\rangle)$, commonly known as the *Bell state* [30]. As a vector, it would be written as $1/\sqrt{2} \cdot [1\,0\,0\,1]^{\mathrm{T}}$. If this state is measured, we have equal probabilities of obtaining as outcome one of the basis states $|00\rangle$ and $|11\rangle$, and zero probability of seeing the states $|01\rangle$ and $|10\rangle$.

In addition to superposition, this quantum state shows *entanglement.* Measuring a value for one qubit of the Bell state would immediately fix the value of the other qubit corresponding to the measurement outcome, e.g., after measuring $q_1 = |0\rangle$ (or $q_1 = |1\rangle$) we immediately know that $q_0 = |0\rangle$ (or $q_0 = |1\rangle$).

Quantum states are manipulated through quantum gates. A quantum gate is any linear operator mapping quantum states to quantum states, i.e., a unitary matrix $U \in \mathbb{C}^{2^n \times 2^n}$, where n is the number of qubits. A quantum algorithm consists of a series of gates applied sequentially, e.g., $U = U_m \cdots U_1$ is an algorithm consisting of m gates, which first applies U_1, then U_2, up to U_m. Thus, U denotes the unitary matrix corresponding to applying all the gates. If a quantum state $|\phi\rangle$

serves as input to U, then the output is the quantum state $U \cdot |\phi\rangle$. We say that a quantum algorithm $U_1, \ldots, U_m$ is equivalent to another quantum algorithm $V_1, \ldots, V_\ell$ iff they effect the same unitary matrix, i.e., if $U_m \cdots U_1 = V_\ell \cdots V_1$.

Example 2. Three examples of common quantum gates are the single-qubit phase-shift operation S, the single-qubit Hadamard operation H, and the two-qubit controlled-NOT operation $CNOT$ (here shown with control on the first qubit, target on the second qubit).

$$S = \begin{bmatrix} 1 & 0 \\ 0 & i \end{bmatrix} \quad H = \frac{1}{\sqrt{2}} \begin{bmatrix} 1 & 1 \\ 1 & -1 \end{bmatrix} \quad CNOT = \begin{bmatrix} 1 & 0 & 0 & 0 \\ 0 & 1 & 0 & 0 \\ 0 & 0 & 0 & 1 \\ 0 & 0 & 1 & 0 \end{bmatrix}$$

The state from Example 1 can be created by starting the $|00\rangle$ basis state and then applying a Hadamard operation on the first qubit, followed by a controlled-NOT. Using the tensor product for parallel composition of gates as usual [30], this can be written as $CNOT \times (H \otimes \mathbb{I}) \times |00\rangle = |11\rangle$. Figure 1 shows another example of a 3-qubit quantum circuit built from these three gates generalized to multiple qubits. The circuit is to be read from left to right, so that the Hadamard gate is the first gate, applied to qubit q_2. The $\bullet$ denotes the control qubit of a $CNOT$ gate; the $\oplus$ denotes its target qubit.

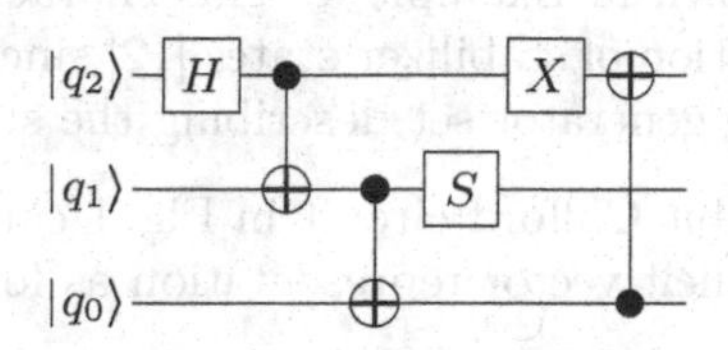

Fig. 1. A quantum circuit on three qubits, q_2–q_0.

An important, though non-universal, subset of quantum circuits are Clifford circuits [14], which consist only of the Clifford gates S, H and $CNOT$. Clifford circuits are ubiquitous in quantum computing because they represent the symmetries that occur in all quantum states albeit they cannot generate arbitrary transformations (hence they are non-universal). As a consequence, they play an essential role in error correction [14,15], entanglement distillation [16] and are used in one-way quantum computing [17]. Clifford circuits are intimately related to the Pauli gate set:

$$\mathbb{I} = \begin{bmatrix} 1 & 0 \\ 0 & 1 \end{bmatrix} \quad X = \begin{bmatrix} 0 & 1 \\ 1 & 0 \end{bmatrix} \quad Y = \begin{bmatrix} 0 & -i \\ i & 0 \end{bmatrix} \quad Z = \begin{bmatrix} 1 & 0 \\ 0 & -1 \end{bmatrix}$$

Example 3. The circuit shown in Fig. 1 is a Clifford circuit, since it only consists of Clifford operations and gates that can be built from Clifford operations. The Pauli gates can be built from Clifford gates, namely as $Z = S^2$ and $X = H \times Z \times H$, and $Y = iXZ$; thus, Pauli gates are Clifford gates.

2.2 Classical Quantum Circuit Simulation

The classical simulation of a quantum circuit is the process of simulating a quantum circuit on a classical binary computer. It is an important task in the context of quantifying the capability of physical quantum computers and in the development of quantum algorithms.

Circuit simulation can be conducted in a straightforward fashion by repeated matrix-vector multiplication. The simulation starts with an initial state and applies the quantum gates one after the other. Each quantum operation is represented by a unitary matrix U_t of dimension $2^n \times 2^n$ and each quantum state by a unit vector $|\phi_t\rangle$ of dimension 2^n. The evolution of a state at time step t is then given by $|\phi_{t+1}\rangle = U_{t+1} |\phi_t\rangle$ (with $|\phi_0\rangle$ commonly being the vector representing the all-zero state, $|\phi_0\rangle = |0 \dots 0\rangle$).

Clifford circuits can be efficiently simulated on a classical computer [13], which is surprising given their importance, but not a contradiction given their non-universality. Starting from the all-zero state $|0^n\rangle$, Clifford circuits only yield so-called *stabilizer states*. An n-qubit stabilizer state $|\phi\rangle$ can be uniquely specified by the set of Pauli operators $G = \pm P_1 \otimes \cdots \otimes P_n$ with $P_i \in \{ \mathbb{I}, X, Y, Z \}$ that stabilize it, i.e., which satisfy $G |\phi\rangle = |\phi\rangle$. This set forms an abelian group, and can always be described succinctly by a set of n generators $G_1, \dots, G_n \in \pm \{ \mathbb{I}, X, Y, Z \}^n$. We can thus think of this generator set as an $n \times n$ matrix of local Pauli operators, where each row (a generator) also has an additional plus or minus sign as shown in Example 4. This characterization is the key to efficiently classical simulation of stabilizer states [12] since the Clifford gates can be applied directly to the generator set describing the state.

Example 4. The three-qubit Clifford circuit in Fig. 1 can be simulated for time steps $t = 0, 1, 2$ using explicit vector representation as follows.

$$
\begin{aligned}
|\phi_0\rangle &= |000\rangle && \text{apply } H \text{ on } q_2 \rightarrow \\
|\phi_1\rangle &= \frac{1}{\sqrt{2}} |000\rangle + \frac{1}{\sqrt{2}} |001\rangle && \text{apply CNOT on control } q_1 \text{ and target } q_2 \rightarrow \\
|\phi_2\rangle &= \frac{1}{\sqrt{2}} |000\rangle + \frac{1}{\sqrt{2}} |011\rangle && \text{(which is } |0\rangle \otimes \text{Bell state)}
\end{aligned}
$$

Alternatively, we may represent each $|\phi_t\rangle$ as a generator set $G(\phi_t) = \{G_1, G_2, G_3\}$.

$$
\begin{aligned}
G(\phi_0) &= \{ Z \mathbb{I} \mathbb{I}, \mathbb{I} Z \mathbb{I}, \mathbb{I} \mathbb{I} Z \} && \text{apply } H \text{ on } q_2 \rightarrow \\
G(\phi_1) &= \{ Z \mathbb{I} \mathbb{I}, \mathbb{I} Z \mathbb{I}, \mathbb{I} \mathbb{I} X \} && \text{apply CNOT on control } q_1 \text{ and target } q_2 \rightarrow \\
G(\phi_2) &= \{ Z \mathbb{I} \mathbb{I}, \mathbb{I} Z Z, \mathbb{I} X X \}
\end{aligned}
$$

We call the generator set representing a stabilizer state a stabilizer tableau. The *stabilizer formalism* stipulates how the tableau should be modified for different Clifford gates [12,13] as exemplified in Example 4. It forms a non-universal, but classically tractable region of quantum computing, whereas decision diagrams considered in this work target universal quantum computing. Nonetheless, Clifford circuits and stabilizer states are important in many domains, as discussed in the introduction.

2.3 Verification of Quantum Circuits

A popular similarity metric for comparing two quantum circuits U, V is the *average fidelity*, $\mathcal{F}_{\text{avg}}(U,V) = \frac{1}{1+2^n}(1 + |\text{tr}(UV^\dagger)|^2)$, where $\text{tr}(M)$ denotes the trace of M. This metric has value 1 iff $U = V$, and has $\mathcal{F}_{\text{avg}}(U,V) < 1$ otherwise. Burgholzer et. al [6], building on a result from Kueng and Gross [32] showed how the average fidelity relates to inputs with random stabilizer states:

Theorem 1 (Burgholzer et. al [6]). *Suppose $|g\rangle$ is a random stabilizer state, and U, V are unitary matrices. Then there is the following relationship between the expectation value and the average fidelity:*

$$\mathbb{E}_{|g\rangle}[\langle g| V^\dagger \cdot U |g\rangle] \approx \mathcal{F}_{\text{avg}}(U,V) \tag{1}$$

Consequently, the average fidelity $\mathcal{F}_{\text{avg}}(U,V)$ can be approximated by simulating the two circuits on several random stabilizer states. Put another way, this quantifies the statement that a random stabilizer state is to a quantum circuit what a random input is to a classical circuit. This motivates the approach of Burgholzer et al.: they repeatedly generate (two copies of) a random stabilizer state $|g\rangle$ as input, then they classically simulate the two circuits on this input, obtaining states $U|g\rangle, V|g\rangle$. Lastly, they compute the inner product of the output states $\langle g| V^\dagger \cdot U |g\rangle$; if the absolute magnitude of this number is smaller than 1, then $|g\rangle$ is a counterexample which certifies that $U \neq V$.

We use this reduction from circuit verification to circuit simulation as the motivation for the setup in our case study in Sect. 5.

3 Motivation

Efficient classical simulation of quantum circuits is an important task for the development of both quantum circuits and their compilation toolchains [7,9]. As reviewed above, this task is conceptually simple (matrix-vector multiplication), but practically hard due to the memory requirements of classical descriptions of quantum states, i.e., state vectors require an exponential amount of memory with respect to the number of qubits. However, certain families of quantum circuits, such as those consisting only of Clifford gates, can be simulated in polynomial time by the stabilizer formalism that exploits the strong algebraic structure present in stabilizer states [12,13]. These techniques have the disadvantage that they do not encompass all of quantum computing, since they cannot produce all quantum gates, i.e., they are not *universal.* For general quantum circuits, i.e., those with no restrictions on the gate set, decision diagrams as reviewed later in this section are a promising data structure to drastically reduce memory requirements in many cases. However, the stabilizer formalism and decision diagrams have thus far excelled only in their respective areas. LIMDDs unite the capabilities of both worlds and thereby enable efficient representation of multiple classes of quantum states.

Although the verification of a given quantum circuit is likely more difficult than simulation of that circuit on a given input, Burgholzer et al. [6]

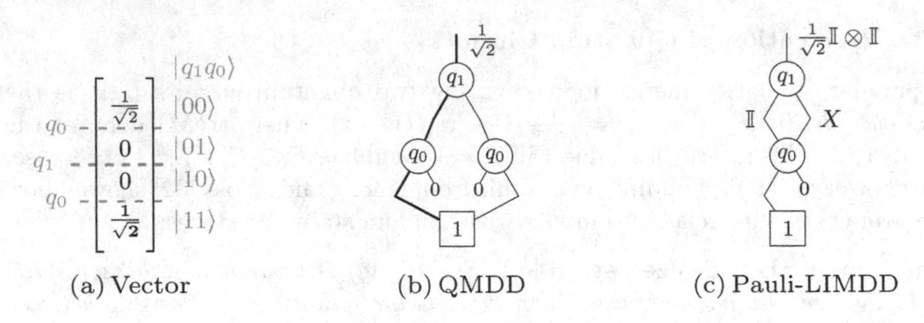

Fig. 2. Different representations of a Bell state

recently showed that simulation can nevertheless be very useful for verification as explained in Sect. 2.3. In particular, they showed that a promising approach is to simulate the circuit on a certain state called a stabilizer state, and gave qualitative and quantitative analytical guarantees on the errors found by this method. Since LIMDDs excel on this family of quantum states, we adopted this approach for our case study in Sect. 5.

The remainder of this section first reviews the basics of QMDDs and LIMDDs, and then discusses their respective strengths (as far as they have been analytically investigated thus far). Based on this, we motivate the need for an implementation of LIMDDs.

3.1 Quantum Multiple-valued Decision Diagrams (QMDDs)

Representing quantum states and operations in a straightforward fashion as vectors and matrices requires an exponential amount of memory with respect to the number of qubits. Decision diagrams are an established data structure that approach this problem by providing a compact representation by exploiting redundancies in the data in many cases. There are multiple types of decision diagrams for the quantum domain [23–27]. We focus on *Quantum Multiple-valued Decision Diagrams* (QMDDs, [26,33]) since they are the state of the art for decision diagram-based quantum circuit simulation.

Conceptually, the QMDD corresponding to the amplitude vector $|\phi\rangle \in \mathbb{C}^{2^n}$ can be built as follows. First, we repeatedly split the amplitude vector in two equal halves, until the individual amplitudes are reached, thus obtaining a binary tree (of height n), in which a node at height k represents a (sub-)vector of length 2^k. Next, whenever two nodes represent states $|\phi\rangle, |\psi\rangle$ satisfying $|\phi\rangle = \lambda \cdot |\psi\rangle$ for some $\lambda \in \mathbb{C}$, we merge these two nodes (discarding one of the subtrees) and place the factor λ on one of the two incoming edges. This *reduced* QMDD is now a directed acyclic graph (DAG) and no longer a tree. Thus, the way the QMDD exploits structure in the vector is by recognizing repeated sub-vectors (or more precisely: sub-vectors which are equal up to a complex constant), which is how it can avoid the exponential blowup in many cases. Note that this construction is explained for illustration purposes only as working with decision diagrams does not at any point require explicitly storing the vector representing a quantum state. For a formal definition we refer to [29].

One can efficiently apply a given operation U to a state $|\phi\rangle$ when both are given as QMDDs. This is done by the MULTIPLY algorithm, which recursively traverses the decision diagrams of U and $|\phi\rangle$ and builds the DD corresponding to the state $|\psi\rangle = U \cdot |\phi\rangle$. We briefly sketch how this algorithm works. First, note that if $|\phi\rangle = |0\rangle \otimes |\phi_0\rangle + |1\rangle \otimes |\phi_1\rangle$, then the DD node representing $|\phi\rangle$ has two children, v_0 and v_1, which represent the amplitude vectors $|v_j\rangle = |\phi_j\rangle$ for $j = 0, 1$. Similarly, the matrix $U = \sum_{ij} |i\rangle\langle j| \otimes U_{ij}$ is represented by a QMDD node with four children v_{ij}, with each v_{ij} representing the submatrix U_{ij}, a quadrant of U. The algorithm first constructs decision diagrams for the states $U_{ij} \cdot |\phi_j\rangle$ using four recursive calls to MULTIPLY(U_{ij}, v_j). Next, it construct decision diagrams representing the states $|\psi_i\rangle = U_{i0}\,|\phi_0\rangle + U_{i1}\,|\phi_1\rangle$ for $i = 0, 1$ using two calls to a procedure ADD implementing addition on QMDDs. Last, it makes a node whose two children are $|\psi_0\rangle$ and $|\psi_1\rangle$, obtaining a node representing $|\psi\rangle = |0\rangle \otimes |\psi_0\rangle + |1\rangle \otimes |\psi_1\rangle = U \cdot |\phi\rangle$, as intended. We use dynamic programming to store the results of all intermediate, recursive calls to MULTIPLY; as is typically done to avoid the exponential-time behavior occurring when all paths in the DAG are considered. In light of this, we remark that in this work we consider matrices U representing a universal gate set that nonetheless each have a small number of nodes that scales as $\mathcal{O}(n)$.

Thus, while it is instructive to consider how a QMDD may be constructed from a given amplitude vector (in the way described above), our algorithms use a more efficient approach, never "expanding" the decision diagram to its amplitude vector, instead working directly on the DD representation of the vectors and matrices. Indeed, this is the primary strength of decision diagrams in general: that they can work on compressed data without decompressing it first.

Example 5. Consider the quantum state $1/\sqrt{2} \cdot (|00\rangle + |11\rangle)$ from Example 1. Figure 2a shows the corresponding vector, with superimposed information on the splitting by qubit (on the left) and the basis states to each amplitude (on the right). Figure 2b shows the same quantum state represented as QMDD.

Retrieving the amplitude of a given quantum state requires traversing the decision diagram and multiplying the edge weights along the way. For readability, edge weights of 1 are omitted; and edges with weight 0 are cut off and represented as stubs. The bolded path in Fig. 2b represents the state $|00\rangle$, and following it gives $1/\sqrt{2} \cdot 1 \cdot 1 = 1/\sqrt{2}$.

While QMDDs enable compact representation of quantum states in many cases, Vinkhuijzen et al. [27] showed they can become exponentially sized for stabilizer states, which can be efficiently simulated classically using the stabilizer formalism as discussed in Sect. 2.1. These states are the intermediate states of circuits consisting of only Clifford gates, thus preventing QMDDs from simulating such circuits efficiently.

3.2 Local Invertible Map Decision Diagrams (LIMDDs)

LIMDDs remove the limitation of QMDDs that cannot efficiently represent every stabilizer state. They can represent each stabilizer state in polynomial space by not just merging nodes that are equivalent up to a scalar, but also those equivalent up to a LIM transformation while retaining universality. A LIM is similar to the stabilizer generator except that it includes an arbitrary scalar.

Definition 1 (Local Invertible Map (LIM), adapted from [27]**).** *An n-qubit* Local Invertible Map *(LIM) is an operator P of the form $P = \lambda P_n \otimes \cdots \otimes P_1$, where the matrices $P_i \in \{ \mathbb{I}, X, Y, Z \}$ are local Pauli matrices and $\lambda \in \mathbb{C} \setminus \{0\}$. An isomorphism between two n-qubit quantum states $|\varphi\rangle, |\psi\rangle$ is a LIM P such that $P|\varphi\rangle = |\psi\rangle$. We then say that $|\varphi\rangle$ is isomorphic to $|\psi\rangle$, denoted $|\varphi\rangle \simeq |\psi\rangle$. Note that isomorphism is an equivalence relation.*

In fact, LIMDDs can efficiently apply most Clifford gates to any quantum state (i.e., even to non-stabilizer states), without increasing the size of the diagram by more than a factor two. LIMDDs extend QMDDs by annotating an edge not only with a complex-valued weight, but also with a series of local Pauli gates represented by the LIMs. This allows LIMDDs to represent all states at least as succinctly as QMDDs and stabilizer tableaus. Additionally, LIMDDs can efficiently represent states which cannot be represented efficiently with either the stabilizer formalism or QMDDs [27], for instance $|T\rangle \otimes |G\rangle$, where $|T\rangle = \frac{1}{\sqrt{2}}(|0\rangle + e^{i\pi/4}|1\rangle)$ and $|G\rangle$ is a stabilizer state.

The interpretation of a LIMDD is similar to that of a QMDD. Each node still corresponds to a complex vector and, when following an edge, the vector given by the child node is still multiplied by the weight on the followed edge. For LIMDDs, rather than only multiplying the vector with a complex scalar, it is now additionally multiplied by the tensor product of the single-qubit gates on the incoming edge. This is illustrated in the following example.

Example 6. Consider again the state $\frac{1}{\sqrt{2}}(|00\rangle + |11\rangle)$ from Example 1. Figure 2c shows the LIMDD of this state. Note that it uses one node less than the corresponding QMDD since it only requires one node q_0 due to the X operation annotated to the left out-edge of q_1. The node labelled q_0 represents the vector $|0\rangle$. From the root, following the left edge from node q_1 gives the vector $1/\sqrt{2} \cdot \mathbb{I}|0\rangle$, while following the right edge gives $1/\sqrt{2} \cdot X|0\rangle = 1/\sqrt{2} \cdot |1\rangle$. Note that correspondence is shown for illustration purposes only as working with decision diagrams does not require explicitly storing the vector at any point.

3.3 The Need for a LIMDD Implementation

As discussed, stabilizer states and Clifford circuits are ubiquitous in many quantum computing algorithms. Moreover, in the context of verifying quantum circuits, stabilizer states serve as good candidates for counterexamples. Therefore, we stand to profit twice from the exponential advantage that LIMDDs promise over existing decision diagrams: first, since stabilizers capture the symmetries present in all quantum states, LIMDDs likely improve universal simulation; second, when we want to verify a quantum circuit by reduction to simulation with a random stabilizer state, the Pauli-LIMDD is guaranteed to efficiently represent at least the initial state, whereas the QMDD is likely exponential [27, Appendix B].

However, the asymptotic advantage of LIMDDs comes with a price. They require both additional memory for bookkeeping the LIMs on edges and additional time for calculating canonical form of nodes. To the best of our knowledge, it is still unknown how this affects the memory and time use in practice because so far an implementation is absent. Existing implementations of decision diagrams in the classical domain [34–38] and the quantum domain [23–28] have shown that translating the concept of a decision diagram into an efficient and usable program or library is far from trivial. LIMDDs are no exception to this rule and come with new challenges regarding the handling of the LIMs in the nodes and edges of the decision diagram.

4 Implementation of LIMDDs

As discussed in the previous sections, LIMDDs scale exponentially better in many cases compared to QMDDs. However, this advantage comes with an increased overhead to keep track of the local invertible maps annotated to nodes and edges in the decision diagram. Efficient management of this additional information is paramount to implement LIMDDs efficiently. Further crucial ingredients for efficient decision diagram implementations are canonicity and dynamic programming. Canonicity ensures that the diagram is never larger than necessary and uniquely represents a quantum state (in QMDDs) or Boolean function (in BDDs). Dynamic programming ensures that manipulation operations, such as gate applications or measurements, take polynomial time in the size of the diagram representing the state. To ensure canonicity, the implementation must put nodes in canonical form, as worked out in [27], and also store them in a corresponding table. To implement dynamic programming, LIMs must be normalized and stored in caches.

This section discusses both established techniques in developing implementations of decision diagrams in Sect. 4.1 as well as new approaches to efficiently handle the LIMs annotated to the edges of LIMDDs in Sect. 4.2. The established and correspondingly adopted techniques include dynamically-sized unique tables, garbage collection, compute tables, as well as indirect storing of complex numbers [28]. While the aforementioned techniques lay a solid foundation for the LIMDDs, they are not sufficient. Efficient approaches to store and manipulate the annotated local invertible maps are required to exploit the potential of LIMDDs.

4.1 Established Techniques

Implementations of various types of (predominantly classical) decision diagrams have been proposed in the last decades [23–27, 34–38]. Over this time, a lot of effort has been put into translating abstract concepts of decision diagrams into concrete instructions that run efficiently on classical computers. Multiple parts and "tricks" of these implementations are reusable for the LIMDD implementation as well. The following list provides a brief description of the most important building blocks:

Unique Tables store the nodes of the decision diagram and enable efficient detection of redundant nodes. These tables are commonly implemented as hash maps storing the nodes with two levels of indirection: the first level gives the qubit (or variable) and the second level is the hash value of the node, which is recursively calculated from the weights of the out-edges and hashes of the respective successors. The subsequent *strong canonical* identifier [39] of a node is the pointer into the memory of the unique tables, enabling access via constant-time de-referencing of the pointer.
When a new node is created, the unique tables are checked for already existing equivalent nodes. If an equivalent node exists, this node is re-used, otherwise the new node is stored in the unique tables.

Compute Tables cache results of operations to implement dynamic programming, avoiding repetitions of the same calculation. Intuitively, the more compact the decision diagram, the more paths (from root to leaf) traverse through the same nodes in the decision diagram. We can avoid processing exponentially many paths by hashing the operands of the recursive operation that traverses these paths in the diagram. The compute tables are implemented as individual hash tables for different operations, where the hash is calculated from the operand nodes.
Additionally, the compute tables are a key concept that enables efficient operations on decision diagrams by enabling dynamic programming. Without them, operations such as multiplication during circuit simulation would always be exponential, since no previous result could be re-used.

Handling of Complex Numbers requires special consideration to ensure that the limited accuracy of floating point numbers does not lead to wrong results. Two key aspects of these considerations are the introduction of a tolerance in the comparison of the components of the complex numbers and storing the components in a dedicated table to exploit the memory address as strong canonical form (with constant-time dereferencing).

Garbage Collection is frequently run to remove entries of the aforementioned tables that are not needed anymore, e.g., after each applied operation in quantum circuit simulation. For each node, a reference count is used to keep track of its state. Upon removal of nodes, garbage collection is also run on the other tables, such as the compute tables, so that no invalid pointers remain in memory—preventing an inconsistent state between the tables and subsequent illegal memory accesses.

The established techniques described above are used in existing packages of decision diagrams for quantum computing. However, LIMDDs require additional functionality to manage the LIMs which are annotated to edges and nodes.

The next section describes the techniques employed to efficiently integrate the information on local invertible maps into the decision diagrams.

4.2 Implementing Local Invertible Maps

Efficient handling of the LIMs annotated to the nodes and edges in the LIMDDs is *the* requirement to actually realize the exponential advantage over QMDDs for certain quantum circuit simulations. Recall that a LIM consists of a complex factor and a Pauli operator $P = P_1 \otimes \cdots \otimes P_n$ with $P_i \in \{\mathbb{I}, X, Y, Z\}$ denoting a local Pauli operator. We call the latter a *Pauli string.*

The proposed LIMDD implementation still provides a strong canonical form for the nodes to ensure canonicity. We implement the canonical form presented in [27], with minor changes described below, which entails, among others, finding the lexicographically minimal Pauli strings for both of a node's outgoing edges, and possibly swapping the two children. The LIMs are stored in a new table to enable constant-time decisions whether two LIMs are equal. Additionally, because the all-identity operator occurs very frequently, we "hardcode" this as null pointer, to prevent many lookups to the LIM table.

In line with existing work [13], we represent a Pauli operator using two bits, so that the operators $\mathbb{I}, X, Y, Z$ are represented by 00, 01, 11, 10, respectively, and a Pauli string of n operators is stored using $2n + 2$ bits, using 2 extra bits to store a scalar factor in $\{\pm 1, \pm i\}$. This enables efficient multiplication of Pauli operators, namely, the product of two LIMs is obtained by XORing their respective bit strings. For QMDDs, the diagram can be traversed by following edges, which is accomplished simply by dereferencing a pointer and multiplying the weight of the considered edge. For LIMDDs, following an edge is slightly more involved, since the local invertible map can affect each level downwards. To keep track of the LIMs to be applied and to avoid creating a new decision diagram for each followed edge, we keep auxillary information about the current LIM during each step of the traversal.

We now briefly list the biggest changes that are required to turn a QMDD package into a LIMDD package.

Putting Nodes in Canonical Form ensures canonicity, which keeps the diagram as small as possible, by allowing nodes representing redundant subvectors to be merged. We use the canonicity scheme for LIMs as proposed in [27]. In this scheme, a node v always has the identity LIM $\mathbb{I}_2^{\otimes n}$ on its 0-edge and a LIM P on its 1-edge, such that P is the lexicographically minimal LIM possible, in the sense that using any smaller LIM results in a node v' which is not Pauli-isomorphic to v. Since the LIM P depends only on the state vector that the node represents, and is minimal in a precise way, this makes the node canonical. Consequently, the diagram will merge two nodes whenever they represent two Pauli-isomorphic subvectors. This minimal P is found by first finding the stabilizer tableaux (see Section 2.2) of v's two children states, which requires time $\mathcal{O}(n^3)$. This approach amortizes the cost of computing canonical LIMs over the entire DD structure; in other words, to construct a canonical LIM for node v, we only need to inspect the stored stabilizer generator sets of v's children (and not their descendants). This step, of constructing these groups, presents the biggest added computational overhead of all changes, namely the time required for making a new node increases from $\mathcal{O}(1)$ to $\mathcal{O}(n^3)$ because of the need to find stabilizer groups. Still, as shown in [27], this overhead enables a asymptotically exponential advantage.

Edge Weight Normalization is part of making the node canonical. We employ the normalization scheme from [40], which differs from the one proposed in [27]. Namely, when choosing the weights α_0, α_1 on the out-edges of a node, we require that $|\alpha_0|^2 + |\alpha_1|^2 = 1$, as opposed to requiring the left-most non-zero edge weight to be normalized to 1 (meaning that α_0 has to be either 0 or 1). This allows us to better take advantage of the existing cache for complex numbers and faster sampling from the decision diagram. Given such numbers α_0, α_1, we have a choice between α_1 and $-\alpha_1$, and we choose the one having nonnegative imaginary part; ties are broken by choosing the one with nonnegative real part. If we choose $-\alpha_1$, then we correct for this by multiplying the LIM on the incoming edge by $Z \otimes \mathbb{I}^{\otimes n-1}$.

The **Operation Cache** for LIMDDs allows us to improve the caching of the ADD operation results to potentially be more succinct and achieve more cache hits. Specifically, we get a cache hit on input $A|v\rangle + B|w\rangle$ whenever a previous call to ADD had input $C|v\rangle + D|w\rangle$ satisfying $A^{-1}B|w\rangle = C^{-1}D|w\rangle$. We implement this using the caching algorithm from [27], namely, on input $A|v\rangle + B|w\rangle$, if the result is edge $E|r\rangle$, then we add $\text{CACHE}[F, v, w] := E|r\rangle$, where F is a canonically chosen LIM determined by A, B, and w.

The **LIM Table** stores the LIMs on the edges and the LIMs which generate a state's stabilizer group, so that common LIMs are shared in the LIMDD, thus reducing the total memory footprint. Multiple LIMDD sub-routines make use of a state's stabilizer group, e.g., for finding canonical edge labels. We choose to construct a set of LIMs which generate a state's stabilizer group as soon as that state's node is created, using the algorithm by [27], and we store these LIMs in the LIM Table. LIMs that are no longer required are identified via reference counting and removed during garbage collection.

We focus on the efficient storing and manipulation of LIMDDs for quantum states and continue to use QMDDs from the original package [28] to represent quantum operations as explained in Sect. 3.2, so that simulation can still be conducted in a fashion similar to [33]. This does not present a limitation of the approach, since QMDDs efficiently represent all gates considered in this work. Namely, we only use single-qubit gates with arbitrary controls, and these for gates the size of the diagram scales as $\mathcal{O}(n)$. Therefore, storage of the quantum state, rather than the matrix, remains the main memory bottleneck in common simulation scenarios.

To ensure the validity of the LIMDD implementation, we performed extensive tests on more than 1700 quantum circuits of varying sizes, parameters and complexity, ensuring that the intermediate state after every gate is the same as that found by the QMDD.

The next sections provides a case study that compares LIMDDs against QMDDs based on the Quantum Fourier Transform, which is an important building block of many quantum algorithms.

5 Case Study

In this section, we provide the results obtained by an experimental case study of the implementation presented in this paper. To this end, we created a complete, open-source LIMDD package in C++ available at https://github.com/cda-tum/ddsim/tree/limdd, based on an existing open-source implementation for QMDDs provided in [28, 41]. The motivation of the case study is an investigation on the extend that the theoretically proven advantage over QMDDs applies in an actual implementation.

In order to demonstrate the efficacy of the resulting implementation and thereby for the first time empirically comparing LIMDDs and QMDDs, we conducted quantum circuit simulation of a circuit which implements the Quantum Fourier Transform (QFT). The QFT is a common subroutine which is used by many quantum algorithms (notably order-finding in Shor's algorithm, phase estimation, and solving the hidden subgroup problem [30]); thus, verifying the correctness of this circuit is a useful step in the compilation toolchain of many quantum algorithms. We consider QFT circuits for various numbers of qubits, $n = 3 \dots 24$. We simulate the QFT on n qubits using a random n-qubit stabilizer state as the input. This stabilizer state is prepared by prepending a random Clifford circuit with $10 \cdot n$ gates, and then simulating from the initial state $|0\rangle^{\otimes n}$; the output of this circuit is a random stabilizer state.

The evaluation was conducted for QMDDs and LIMDDs on a server running GNU/Linux and GCC-10.3.0 with an AMD Ryzen 9 3950X running at 3.5 GHz and 128 GiB memory.

The results in Fig. 3 show that LIMDDs outperform QMDDs on large instances (from $n = 19$ qubits and up), whereas QMDDs outperform LIMDDs on small instances (up to and including 18 qubits). This is most pronounced at 24 qubits, where LIMDDs are about five times faster. These results are expected: LIMDDs are proven to be asymptotically faster, but this comes at the price of adding a lot of computational overhead in the handling of the LIMs on the edges, as explained in Sect. 4. The data show that this overhead pays off in the long run where the asymptotically better performance becomes realized in practice. In the graph, this translates into the fact that the LIMDD line is less steep than the QMDD line. The LIMDD is still small (it has $\mathcal{O}(n)$ nodes) when it finishes preparing a random stabilizer state and starts simulating the QFT, whereas the QMDD is already very large (it has $2^{\mathcal{O}(n)}$ nodes) at this point. At the end of the QFT, both types of decision diagrams are almost fully populated (i.e., almost of maximum size), since the state after the QFT does not possess much redundancy to be exploited. Generally, applying a gate to a small decision diagram

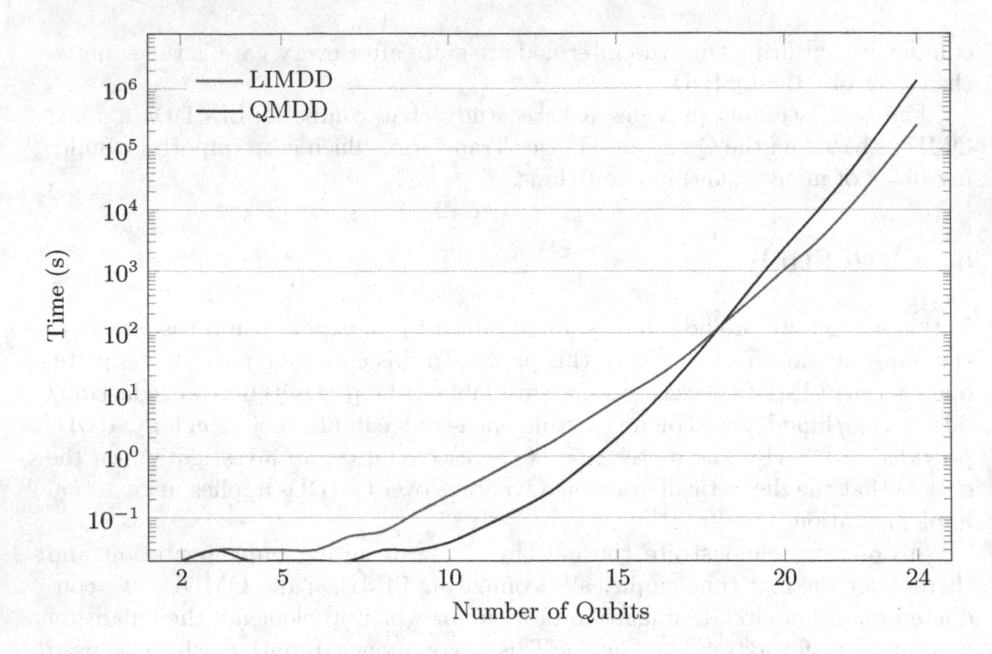

Fig. 3. Quantum Fourier Transform Simulation on Random Stabilizer States

is more efficient than applying the same gate to a large decision diagram, so especially the first few gates of the QFT can be applied quickly by LIMDDs, which eventually leads to the better runtime. In summary, while the additional overhead of LIMDDs outweights the lower complexity for small circuits, they can demonstrate their advantage as the circuit size increases.

6 Conclusions

In this paper, we presented the first implementation of *Local Invertible Map Decision Diagrams* (LIMDDs). The implementation includes techniques adapted from other decision diagram packages (both classical and quantum) that are tried and tested, as well as new considerations to efficiently handle *Local Invertible Maps* (LIMs). By this, we enable the potential of LIMDDs to be realized in practice. A case study confirm that LIMDDs provide an advantage for the classical simulation of quantum circuits that exceed a certain complexity, as shown by the Quantum Fourier Transform. The resulting open-source C++ implementation is available under the MIT license via https://github.com/cda-tum/ddsim/tree/limdd.

Acknowledgment. This work received funding from the European Research Council (ERC) under the European Union's Horizon 2020 research and innovation programme (grant agreement No. 101001318) and was part of the Munich Quantum Valley, which is supported by the Bavarian state government with funds from the Hightech Agenda Bayern Plus.

References

1. Grover, L.K.: A fast quantum mechanical algorithm for database search. In: Symposium on Theory of Computing, pp. 212–219 (1996). https://doi.org/10.1145/237814.237866
2. Montanaro, A.: Quantum-walk speedup of backtracking algorithms. Theor. Comput. **14**(1), 1–24 (2018)
3. Ambainis, A., Gilyén, A., Jeffery, S., Kokainis, M.: Quadratic speedup for finding marked vertices by quantum walks. In: Symposium on Theory of Computing, pp. 412–424 (2020)
4. Shor, P.W.: Polynomial-time algorithms for prime factorization and discrete logarithms on a quantum computer. SIAM J. Comp. **26**(5), 1484–1509 (1997). https://doi.org/10.1137/S0097539795293172
5. Lanyon, B.P., et al.: Towards quantum chemistry on a quantum computer. Nat. Chem. **2**(2), 106 (2010)
6. Burgholzer, L., Kueng, R., Wille, R.: Random stimuli generation for the verification of quantum circuits. In: Asia and South Pacific Design Automation Conference, pp. 767–772, New York, NY, USA,: Association for Computing Machinery. ISBN 9781450379991 (2021)
7. Burgholzer, L., Wille, R.: Advanced equivalence checking for quantum circuits. IEEE Trans. on CAD Integr. Circ. Sys., **40**(9):1810–1824 (2021). https://doi.org/10.1109/TCAD.2020.3032630
8. Burgholzer, L., Raymond, R., Wille, R.: Verifying results of the IBM Qiskit quantum circuit compilation flow. In: 2020 IEEE International Conference on Quantum Computing and Engineering (QCE), pp. 356–365. IEEE (2020)
9. Carette, J., Ortiz, G., Sabry, A.: Symbolic execution of hadamard-toffoli quantum circuits. In: Proceedings of the 2023 ACM SIGPLAN International Workshop on Partial Evaluation and Program Manipulation, pp. 14–26 (2023)
10. Guerreschi, G.G., Matsuura, A.Y.: Qaoa for max-cut requires hundreds of qubits for quantum speed-up. Scientific Reports, **9**(1), 6903 (2019). ISSN 2045-2322. https://doi.org/10.1038/s41598-019-43176-9
11. Jones, T., Brown, A., Bush, I., Benjamin, S.C.: Quest and high performance simulation of quantum computers. Scientific Reports, **9**(1), 10736 (2019). ISSN 2045–2322. https://doi.org/10.1038/s41598-019-47174-9
12. Gottesman, D.: Stabilizer codes and quantum error correction (1997)
13. Aaronson, S., Gottesman, D.: Improved simulation of stabilizer circuits. Phys. Rev. A **70**(5), 052328 (2004)
14. Gottesman, D.: Theory of fault-tolerant quantum computation. Phys. Rev. A **57**, 127–137 (1998)
15. Gottesman, D.: Stabilizer codes and quantum error correction. California Institute of Technology (1997)
16. Bennett, C.H., Bernstein, H.J., Popescu, S., Schumacher, B.: Concentrating partial entanglement by local operations. Phys. Rev. A, **53**, (1996). https://doi.org/10.1103/PhysRevA.53.2046
17. Browne, D., Briegel, H.: One-way quantum computation. Quantum information: From foundations to quantum technology applications, pp. 449–473 (2016)
18. Dijk, T.M., Wille, R., Meolic, R.: Tagged BDDs: Combining reduction rules from different decision diagram types. In: Stewart, D., Weissenbacher, G., editors, Formal Methods in CAD, 2017. https://doi.org/10.23919/FMCAD.2017.8102248

19. Minato, S.: Zero-suppressed BDDs for set manipulation in combinational problems. In: Design Automation Conference, pp. 272–277 (1993)
20. Bryant, R.E.: Symbolic manipulation of Boolean functions using a graphical representation. In: Design Automation Conference, pp. 688–694 (1985)
21. Bryant, R.E., Chen, Y.A.: Verification of arithmetic circuits with binary moment diagrams. In: Design Automation Conference, pp. 535–541 (1995)
22. Drechsler, R., Sarabi, A., Theobald, M., Becker, B., Perkowski, M.A.: Efficient representation and manipulation of switching functions based on Ordered Kronecker Functional Decision Diagrams. In: Lorenzetti, M.J., editor, Design Automation Conference, pp. 415–419 (1994). https://doi.org/10.1145/196244.196444
23. Abdollahi, A., Pedram, M.: Analysis and synthesis of quantum circuits by using quantum decision diagrams. In: Design, Automation and Test in Europe, pp. 317–322 (2006)
24. Wang, S.-A., Lu, C.-Y., Tsai, I.-M., Kuo, S.-Y.: An XQDD-based verification method for quantum circuits. IEICE Trans. Fundamentals, 91-A(2), 584–594 (2008)
25. Viamontes, G.F., Markov, I.L., Hayes J.P.: Quantum Circuit Simulation. Springer (2009). ISBN 978-90-481-3064-1. https://doi.org/10.1007/978-90-481-3065-8
26. Niemann, P., Wille, R., Miller, D.M., Thornton, M.A., Drechsler, R.: QMDDs: Efficient quantum function representation and manipulation. IEEE Trans. on CAD of Integr. Circ. Sys. 35(1), 86–99 (2016). https://doi.org/10.1109/TCAD.2015.2459034
27. Vinkhuijzen, L., Coopmans, T., Elkouss, D., Dunjko, V., Laarman, A.: LIMDD: A decision diagram for simulation of quantum computing including stabilizer states. CoRR, abs/2108.00931, 2021. arxiv.org/abs/2108.00931
28. Zulehner, A., Hillmich, S., Wille, R.: How to efficiently handle complex values? Implementing decision diagrams for quantum computing. In: David Z. Pan, editor, International Conference on CAD, pp. 1–7, 2019. https://doi.org/10.1109/ICCAD45719.2019.8942057
29. Miller, D.M., Thornton, M.A.: QMDD: A decision diagram structure for reversible and quantum circuits. In: 36th International Symposium on Multiple-Valued Logic (ISMVL'06), pp. 30–30. IEEE (2006)
30. Nielsen, M.A., Chuang, I.L.: Quantum Computation and Quantum Information (10th Anniversary edition). Cambridge University Press (2016). ISBN 978-1-10-700217-3. www.cambridge.org/de/academic/subjects/physics/quantum-physics-quantum-information-and-quantum-computation/quantum-computation-and-quantum-information-10th-anniversary-edition?format=HB
31. Jozsa, R.: Quantum algorithms and the fourier transform. Royal Society London. Series A **454**(1969), 323–337 (1998)
32. Kueng, R., Gross, D.: Qubit stabilizer states are complex projective 3-designs. arXiv preprint arXiv:1510.02767 (2015)
33. Zulehner, A., Wille, R.: Advanced simulation of quantum computations. IEEE Trans. on CAD of Integr. Circ. and Sys. 38(5), 848–859 (2019). https://doi.org/10.1109/TCAD.2018.2834427
34. Somenzi, F.: CUDD: CU decision diagram package release 3.0.0. http://www.vlsi.colorado.edu/~fabio/
35. Van Dijk, T., Laarman, A., Van De Pol, J.: Multi-core BDD operations for symbolic reachability. Electronic Notes Theor. Comput. Sci. **296**, 127–143 (2013)
36. Lv, G., Chen, Y., Feng, Y., Chen, Q.L., Su, K.: A succinct and efficient implementation of a 2^{32} BDD package. In: Margaria, T., Qiu, Z., Yang, H., eds, International Symposium on Theoretical Aspects of Software Engineering, pp. 241–244, 2012. https://doi.org/10.1109/TASE.2012.22

37. Herbstritt, M.: wld: A C++ library for decision diagrams. http://www.ira.informatik.uni-freiburg.de/software/wld/ (2004)
38. Knuth, D.E.: The art of computer programming: Binary decision diagrams. http://www-cs-faculty.stanford.edu/knuth/programs.html (2011)
39. Brace, K.S., Rudell, R.L., Bryant, R.E.: Efficient implementation of a BDD package. In: Smith, R.C., editor, Design Automation Conference, pp. 40–45, 1990. https://doi.org/10.1145/123186.123222
40. Hillmich, S., Markov, I.L., Wille, R.: Just like the real thing: Fast weak simulation of quantum computation. In: Design Automation Conference, pp. 1–6. IEEE (2020). https://doi.org/10.1109/DAC18072.2020.9218555
41. Wille, R., Hillmich, S., Burgholzer, L.: JKQ: JKU tools for quantum computing. In Int'l Conference on CAD, pp. 154:1–154:5 (2020). https://doi.org/10.1145/3400302.3415746

ParaGnosis: A Tool for Parallel Knowledge Compilation

Giso H. Dal[1(✉)], Alfons Laarman[2], and Peter J. F. Lucas[1]

[1] Faculty of Electrical Engineering, Mathematics and Computer Science, University of Twente, Enschede, The Netherlands
{g.h.dal,peter.lucas}@utwente.nl

[2] Leiden Institute of Advanced Computer Science, Leiden University, Leiden, The Netherlands
a.w.laarman@liacs.leidenuniv.nl

Abstract. ParaGnosis (https://doi.org/10.5281/zenodo.7312034, https://zenodo.org/badge/latestdoi/560170574, Alternative url: https://github.com/gisodal/paragnosis, Demo url: https://github.com/gisodal/paragnosis/blob/main/DEMO.md) is an open-source tool that supports inference queries on Bayesian networks through weighted model counting. In the knowledge compilation step, the input Bayesian network is *encoded* as propositional logic and then *compiled* into a knowledge base in decision diagram representation. The tool supports various diagram formats, including the Weighted-Positive Binary Decision Diagram (WPBDD) which can concisely represent discrete probability distributions.

Once compiled, the probabilistic knowledge base can be queried in the inference step. To efficiently implement both steps, ParaGnosis uses simulated annealing to split the knowledge base into a number of partitions. This further reduces the decision diagram size and crucially enables parallelism in both the compilation and the inference steps. Experiments demonstrate that this partitioned approach, in combination with the WPBDD representation, can outperform other approaches in the knowledge compilation step, at the cost of slightly more expensive inference queries. Additionally, the tool can attain 15-fold parallel speedups using 64 cores.

1 Introduction

Hazard and safety analysis are important tools to mitigate risks and prevent disasters in many industries. The complicated interactions between industrial processes and production chains are often modeled in fault trees [31], Bayesian networks [26], and other graphical models [15], that support reasoning under uncertainty. Probabilistic reasoning is however computationally hard. To remedy this problem, *knowledge compilation* aims to find a concise representation that supports getting fast results on queries regarding the same probabilistic model.

Many representation languages [5,14,22,29,32] have been studied for this purpose, analytically as well as experimentally, demonstrating a clear tradeoff

G. Caltais and C. Schilling (Eds.): SPIN 2023, LNCS 13872, pp. 22–37, 2023.
https://doi.org/10.1007/978-3-031-32157-3_2

between the succinctness of the language —with often exponential separations— and the tractability of important operations on them. These operations can roughly be divided into *manipulation operations* and *queries*. The former plays an important role in the compilation step, which builds the knowledge base, while the latter are used to query it. For this reason, the budget for manipulation operations is often greater, since this step needs to happen only once.

To the best of our knowledge, ParaGnosis is the first parallel knowledge compilation and inference tool for Bayesian networks with discrete variables. It compiles networks into the Weighted-Positive Binary Decision Diagram (WPBDD). Our chosen parallelization approach through partitioning can reduce the effort spent on compilation because the decomposition of a propositional theory is known to yield smaller symbolic representations, which has previously been shown in model checking [15,17,25,28]. Empirical results with ParaGnosis confirm this [13]. This improvement is offset by a potential increase in the time spent on inference, although our experiments still demonstrate good performance due to the smaller representations. The parallelization approach is orthogonal with regard to target representation languages [11], as we demonstrated with four different representations.

The performance of ParaGnosis compares favorably [10,11] against other knowledge compilers, like SDD, CUDD and ACE, which target SDD [14], OBDD [3] and d-DNNF [16], respectively. The scalability of the tool is good for larger networks and for both compilation and inference, exhibiting over tenfold speedups. ParaGnosis achieves this through its unique compositional approach and use of parallelism.

In this paper, we present the tool ParaGnosis. To help readers understand how the tool works, a high-level overview of the theoretical concepts is given (Sect. 2 and 3) that underly its implementation (Sect. 4). We only offer a user-oriented description of the used BDD, and partitioning and parallelization concepts (see [10,11] for in-depth descriptions and core algorithms). Performance results are presented in Sect. 5. Some examples, figures and definitions are borrowed from previously published works that describe the theoretical foundation of ParaGnosis [10,12]. We finally discuss how this tool relates to others in its field (Sect. 6). Compared to previous work, we have improved ParaGnosis to handle queries other than marginalization, including conditional probabilities and automatic posterior computations of all unobserved variables. We also added several inference query optimizations through parallelism, and a graphical visualization is provided of compiled representations.

2 Background

2.1 Bayesian Networks

A Bayesian network (BN) $\mathcal{B}$ is a probabilistic graphical model that represents a joint probability distribution over its variables. Let $X = \{X_1, \ldots, X_n\}$ be a set of random variables.

Values of a variable X_1 are denoted in lowercase. We denote with $P(X = x)$ the probability that $(X_1, \ldots, X_n) = (x_1, \ldots, x_n)$, i.e. $X_i = x_i$, for $i = 1, \ldots, n$. Let $I \subseteq [n]$, then $X_I = \{X_i \mid i \in I, X_i \in X\}$.

Definition 1 (Bayesian Networks). *A Bayesian network $\mathcal{B} = (\mathcal{G}, P)$ is a DAG $\mathcal{G} = (V, E)$, with nodes V and edges $E \subseteq V \times V$, that models a factorization of joint probability distribution $P(X_V)$ defined over random variables X_V as:*

$$P(X_V = x_V) = \prod_{v \in V} P(X_v = x_v \mid X_{\mathsf{pa}(v)} = x_{\mathsf{pa}(v)}), \tag{1}$$

such that there is a one-to-one correspondence between nodes V and variables X_V, and the conditional probability distribution of $X_v \in X_V$ given its parents $X_{\mathsf{pa}(v)}$ is specified as $P(X_v \mid X_{\mathsf{pa}(v)})$. A Conditional Probability Table (CPT) displays the conditional probabilities $P(X_v \mid X_{\mathsf{pa}(v)})$ of a single variable X_v with respect to its mutually dependent random variables $X_{\mathsf{pa}(v)}$.

Example 1 (Bayesian Network). Fig. 1 shows a BN $\mathcal{B}$ defined over variables $X = \{A, B\}$ (Fig. 1b), its CPTs (Fig. 1a)

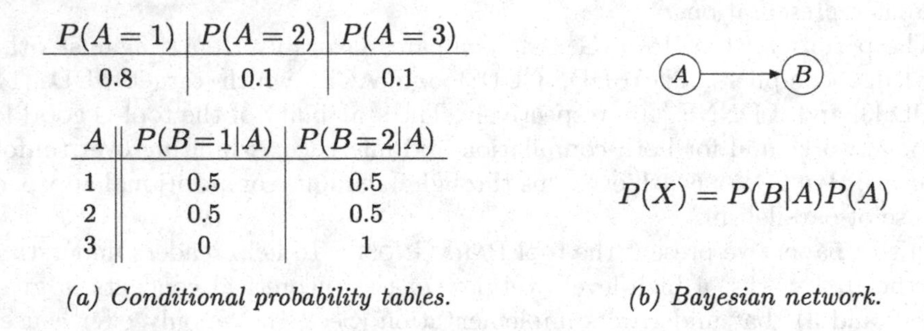

$P(A=1)$	$P(A=2)$	$P(A=3)$
0.8	0.1	0.1

A	$P(B=1 \vert A)$	$P(B=2 \vert A)$
1	0.5	0.5
2	0.5	0.5
3	0	1

(a) Conditional probability tables. *(b) Bayesian network.*

Fig. 1. Bayesian network with local structure.

Posterior probabilities can be computed —a process called inference— using well-known lemmas in probability theory, such as Bayes' theorem $P(X|Y) = \frac{P(X|Y)P(Y)}{P(Y)}$, marginalization $P(X) = \sum_y P(X, Y = y)$, and the factorization property (Definition 1).

2.2 Knowledge Compilation

Bayesian networks represent concise factorizations of probability distributions by using conditional independence assumptions. The size of the factorization has direct implications toward the cost of reasoning, i.e., probabilistic inference. A more expressive model must be used to further improve a BN's factorization in order to exploit additional independences [5]. A prominent way of achieving this is to find a more concise and canonical representation, called a *knowledge base*, such as a Binary Decision Diagram (BDD) [3]. Compiling a BN into a decision diagram (DD) representation is commonly referred to as *knowledge compilation* [16], or simply compilation.

Encoding. BNs are defined over multi-valued domains. Prior to compiling it to a DD, we require an encoding to transition from the multi-valued domain to the Boolean domain. There are multiple ways to do this. We choose to first translate a BN into a Boolean formula with dedicated variables to represent probabilities [5,12].

Conjunctive Normal Form (CNF) is commonly used to facilitate compilation. This CNF is constructed as follows. We create for every $X_v \in X$ a set of atoms $\mathsf{at}(X_i) = \{x_1, ..., x_n\}$. Semantically, $x_i \in \mathsf{at}(X_v)$ represents X_v being equal its ith value. An atom ω_j is introduced for every unique probability in X_v's CPT, i.e., ω_j can refer to multiple distinct entries in X_v's CPT if they represent the same probability. This associated probability is obtained by $\mathsf{pr}(\omega_i) \in [0, 1]$. Function pr returns 1 if no probability is associated with its argument. Finally, a clause is created for each unique valuation of X_v CPT (e.g. for each probability) disjoined with a probability associated ω_j that belongs to that valuation. Also, clauses are added to prevent inconsistent valuation representations, e.g., a variable having multiple values at the same time. We now show this by example, but detailed descriptions can be found in [12]

Example 2 (Bayesian Network encoding).
Let BN $\mathcal{B} = (\mathcal{G}, P)$, with $\mathcal{G} = (V, E)$, be defined over variables $X_V = \{A, B\}$ as described in Example 1. For simplicity, we will focus on just variable A. To encode the BN we create atoms $\mathsf{at}(A) = \{a_1, a_2, a_3\}$. A's CPT has 3 distinct entries and only two distinct probabilities. We create ω_1 for valuations $A = 1$ and create ω_2 for $A = 2$ and $A = 3$, with $\mathsf{pr}(\omega_1) = 0.8$, $\mathsf{pr}(\omega_2) = 0.1$.

The CNF representation follows:

$$(a_1 \vee a_2 \vee a_3) \wedge (\overline{a_1} \vee \overline{a_2}) \wedge (\overline{a_1} \vee \overline{a_3}) \wedge (\overline{a_2} \vee \overline{a_3}) \wedge$$
$$(\overline{a_1} \vee \omega_1) \wedge (\overline{a_2} \vee \omega_2) \wedge (\overline{a_3} \vee \omega_2)$$

The first row is solely concerned with preserving valuation consistency (A can only have one value). The second row has a clause for each valuation/probability. The encoding includes the following models for variable A:

	Models					Weights
1	a_1	$\overline{a_2}$	$\overline{a_3}$	ω_1	$\overline{\omega_2}$	$\mathsf{pr}(\omega_1) \cdot \mathsf{pr}(\overline{\omega_2}) = 0.8 \cdot 1 = 0.8$
2	$\overline{a_1}$	a_2	$\overline{a_3}$	$\overline{\omega_1}$	ω_2	$\mathsf{pr}(\overline{\omega_1}) \cdot \mathsf{pr}(\omega_2) = 1 \cdot 0.1 = 0.1$
3	$\overline{a_1}$	$\overline{a_2}$	a_3	$\overline{\omega_1}$	ω_2	$\mathsf{pr}(\overline{\omega_1}) \cdot \mathsf{pr}(\omega_2) = 1 \cdot 0.1 = 0.1$

Note that the weighted model count sums to 1.0 for this selection of models. However, there are more models, e.g., model a_1,$\overline{a_2}$,$\overline{a_3}$,ω_1,ω_2, model $\overline{a_1}$,a_2,$\overline{a_3}$,ω_1,ω_2, etc. Only minimal models sum to 1.0, i.e., models with the most amount of negations.

Compilation. Now that we have an encoding, we can look at its compilation to a WPBDD in particular. A WPBDD is an ordered BDD that represents a concise factorization of a Boolean formula f as a (rooted) directed acyclic graph with decision nodes, and two terminal nodes labeled with 1 and 0. Each non-terminal node v is labeled with a Boolean variable $\mathsf{var}(v) = x_v$ and has two children, $\mathsf{hi}(v)$ and $\mathsf{lo}(v)$, with a set of weight variables $\mathsf{wg}(v)$ at the edge to node $\mathsf{hi}(v)$. Edges to nodes $\mathsf{hi}(v)$ and $\mathsf{lo}(v)$ are solid and dotted, respectively, as shown in Fig. 2a. Its logical equivalent is shown in Fig. 2b. Each root-terminal path contains a variable at most once, and in a particular total or partial order.

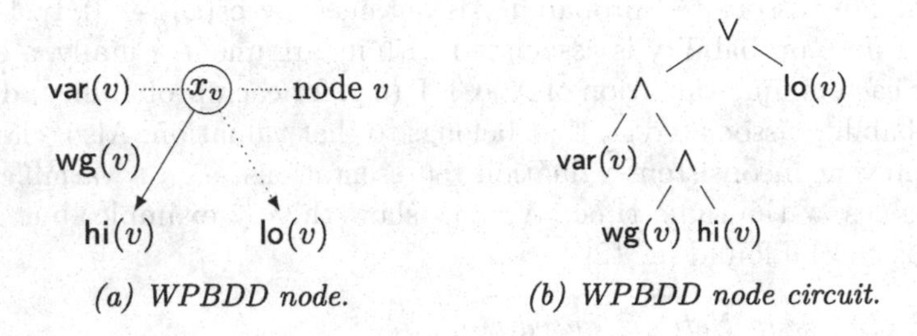

(a) WPBDD node. *(b) WPBDD node circuit.*

Fig. 2. The semantics of a WPBDD node.

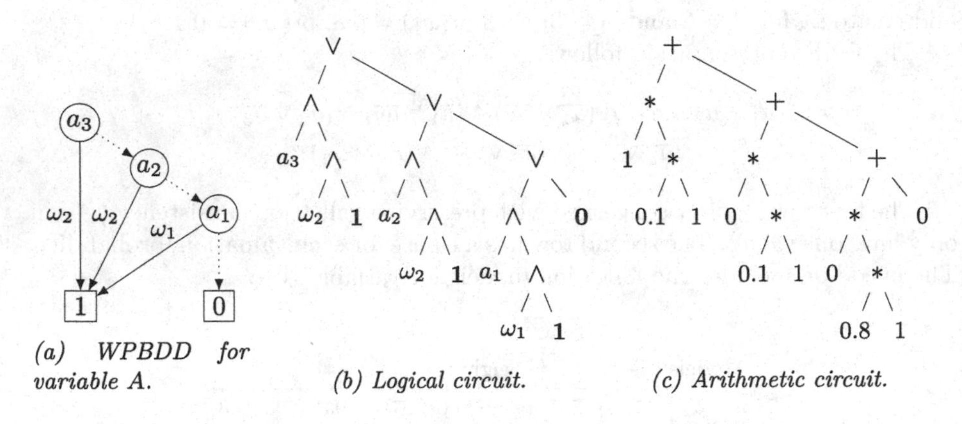

(a) WPBDD for variable A. *(b) Logical circuit.* *(c) Arithmetic circuit.*

Fig. 3. Performing inference in Example 3.

A CNF encoding as described above acts as an entry point for the language compiler [20]. Such compilers target different variations of DDs.

The respective DD is built using the typical bottom-up strategy [3], by applying DD operations to construct a DD representing the encoded formula from the previous step. The process of compiling into a respective DD is by far the most expensive operation, compared to the inference step, which is linear in the size of the knowledge base as desired.

Inference. Inference is performed through *Weighted Model Counting* on the DD, WMC for short [5,17]. This process sums the weight of every variable assignment. In the decision diagram, these assignments are represented by paths and the weights by edge labels (see the example WPBDD in Fig. 1 (c)). Since these paths often overlap in the DD structure, inference through model counting is linear in the size of the target representation [16].

Let's look at a WPBDD compilation and inference example. A WPBDD exactly represents the encoding provided. In order to perform inference we can trivially transform the logical circuit that the WPBDD represents into an arithmetic circuit.

Example 3 (Compilation and inference). Consider only variable A in Example 1. The compiled representation is shown in Fig. 3a for variable ordering $a_3 \prec a_2 \prec a1$. We have not optimized the representation in order to make the upcoming discussion easier. Reduction rules specific to WPBDDs allow the removal of the a_2 node to further reduce its size. Each path from the root to the **1**-terminal semantically implies evidence. There are three possible paths shown below. If we have evidence prior to traversing the compiled representation, we only consider the paths that are consistent with the evidence.

Path	Logic	Semantics
$a_3 \rightarrow \mathbf{1}$	$\overline{a_1} \wedge \overline{a_2} \wedge a_3$	$A = 3$
$a_3 \dashrightarrow a_2 \rightarrow \mathbf{1}$	$\overline{a_1} \wedge a_2 \wedge \overline{a_3}$	$A = 2$
$a_3 \dashrightarrow a_2 \dashrightarrow a_1 \rightarrow \mathbf{1}$	$a_1 \wedge \overline{a_2} \wedge \overline{a_3}$	$A = 1$

The underlying logical circuit is shown in Fig. 3b (obtained with the circuit in Fig. 2b). To perform inference, we need to instantiate the equivalent arithmetic circuit. Figure 3c shows the instantiated circuit that allows us to compute $P(A = 3) = 0.1$, by setting valuation $(a_1, a_2, a_3) = (0, 0, 1)$.

3 Weighted Model Counting Methodologies

We distinguish between two methodologies within the field of probabilistic inference through WMC. Traditional (or monolithic) Weighted Model Counting and Compositional Weight Model Counting. The former is described in Sect. 2. The latter is introduced by ParaGnosis, and expands traditional compilation with partitioning. The purpose of partitioning is to alleviate high compilation costs. Figure 4 shows an overview of the underlying principles behind the two methodologies, divided into two steps: *compilation* and *inference*. On the left, we have the monolithic and on the right, we have the compositional approach.

In-depth descriptions of compositional WMC in scrupulous detail can be found in [10], however, we consider it to be out of scope for this article and will proceed to give a more high-level description here. The concrete implementation of ParaGnosis is described in Sect. 4.

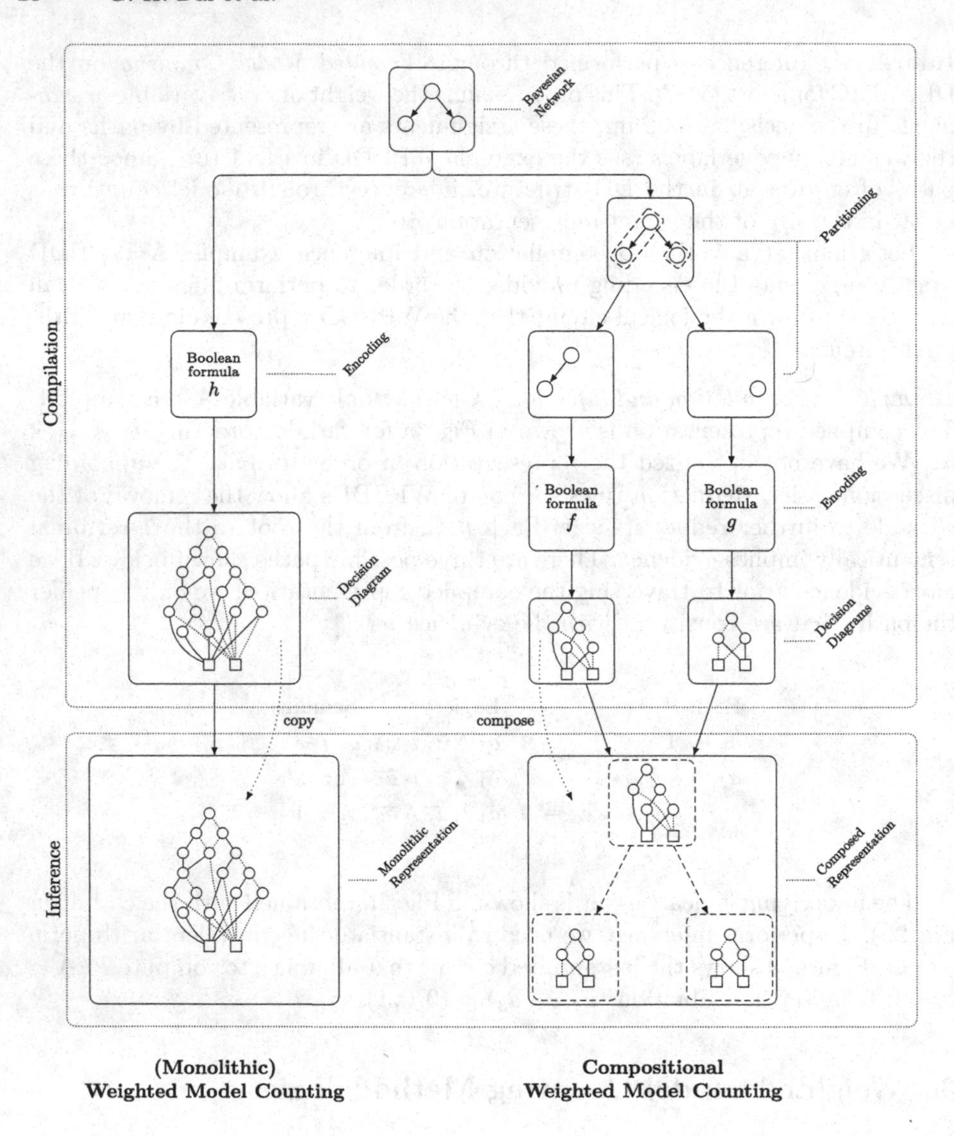

Fig. 4. Compilation and inference process. ParaGnosis uses the compositional approach depicted on the right.

3.1 Compilation to a Compositional Knowledge Base

We first describe compilation as all steps required to obtain a DD. In addition to traditional compilation, ParaGnosis introduces partitioning to further improve overall performance [13]. It finds a partitioning that decomposes the BN into independent components. A component is a set of nodes that are not connected to nodes outside of the component by removing edges from the BN. The fewer removals the better, with regard to inference complexity. This method thus keeps

CPTs intact, as the partitioning happens on a BN node level, not deeper. We give a partitioning demonstration later, using Example 4. With the partitioning in hand, the following steps can be performed independently, per partition. Each partition is considered an independent BN from this point on.

With monolithic compilation, we would only be able to amortize the cost of compilation by performing many inference queries. With partitioned compilation, we shift some of this cost over to the inference side, yielding overall performance improvements in cases where we would traditionally not be able to compile a knowledge base due to time or memory restrictions [13].

3.2 Inference

After compilation, we arrive at the inference step.

In case the user chose to partition the BN during the compilation step, ParaGnosis performs an adapted WMC step to recombine the compiled DDs. A partition's representation should be connected to another when they share a common variable. This implies that we need to traverse partitions according to some partial order. It can be represented by a tree, which we suitably refer to as a composition-tree [10].

The order in which we choose to traverse partitions (combined with common variable relations) determines how they are connected. As we traverse one partition, its sink is connected to the next partition's root. Now that all partitions form one connected component, we can proceed as previously described with a traversal we are already familiar with from WMC. We discuss how partitioning influence inference complexity in Sect. 6.

Example 4 (Partitioned compilation and composition). Consider the BN from Example 1. Figure 5a shows a partitioning. Each partitioning can then independently be compiled to a target representation, i.e. a WPBDD, OBDD, SDD, etc. Figure 5b and 5c show the WPBDD representations of the partitions. Note that partition $\mathcal{G}_2$ also includes $\mathtt{at}(A) = \{a_1, a_2, a_3\}$ in its compiled representation, because B depends on A in the BN. Figure 5d shows how the compiled representations are connected to form one connected component. This facilitates its traversal as a monolithic structure and enables WMC.

4 ParaGnosis

We present the overview of ParaGnosis's architecture in Fig. 6, including its inputs and outputs. It implements the principles outlined in Fig. 4 and Sect. 3. ParaGnosis is a collection of two applications written in C++. The compilation step is implemented by the `COMPILER`, whereas the inference step is implemented by the INFERENCE ENGINE. We provide a high-level view of the implementations in the following sections. In-depth descriptions on the used partitioning technique are provided in [10,13], and how they created independencies are exploited through parallelism in [11].

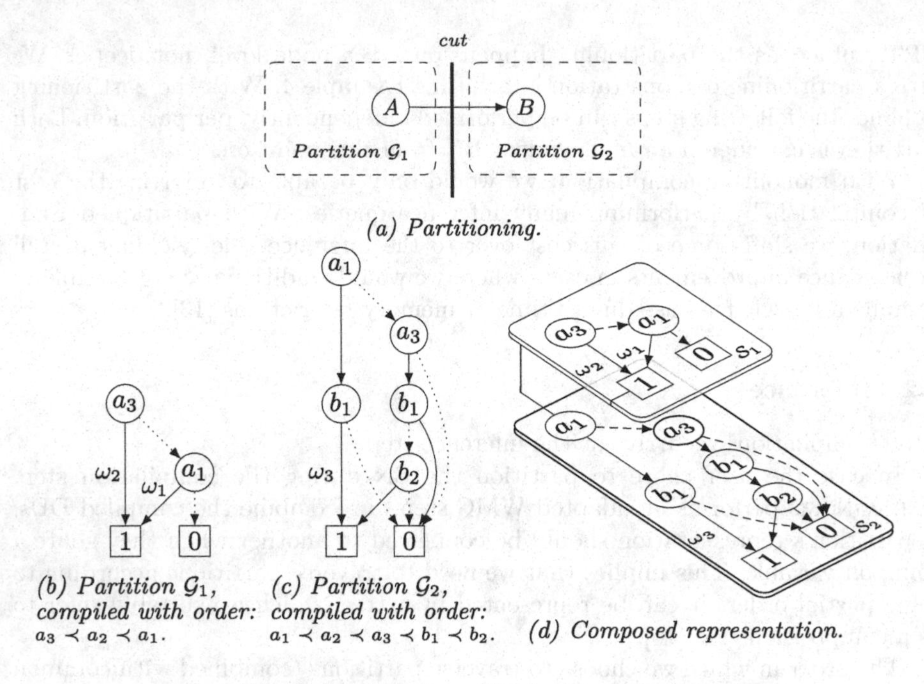

(a) *Partitioning.*

(b) *Partition $\mathcal{G}_1$, compiled with order:* $a_3 \prec a_2 \prec a_1$.

(c) *Partition $\mathcal{G}_2$, compiled with order:* $a_1 \prec a_2 \prec a_3 \prec b_1 \prec b_2$.

(d) *Composed representation.*

Fig. 5. Partitioning a BN for Example 4.

ParaGnosis comes with a wrapper script called pg, simplifying compilation and inference steps. Compiling the 'icy roads' network to a WPBDD and directly visualizing it (using DOT) is as easy as running the following command:

```
> pg compile icy_roads --method wpbdd --dot
```

Computing posteriors for variables 'Holmes' and 'Icy' can be achieved by:

```
> pg inference --posteriors="Holmes,Icy" icy_roads
```

An extended tool demo is also available[1].

4.1 The COMPILER

As shown in Fig. 6, the COMPILER takes as input a discrete BN in the original HUGIN's `.net` format [23]. The compiler is responsible for creating a decision diagram from the provided BN and writes these to output files. The principles outlined in Sect. 3 provide an orthogonal framework with regard to the target representation. However, ParaGnosis chooses to target *Weighted Positive Binary Decision Diagrams* (WPBDD), because it is a dedicated representation for probabilistic inference. (We discuss differences with other representations in Sect. 6.)

[1] Demo: https://github.com/gisodal/paragnosis/blob/main/DEMO.md.

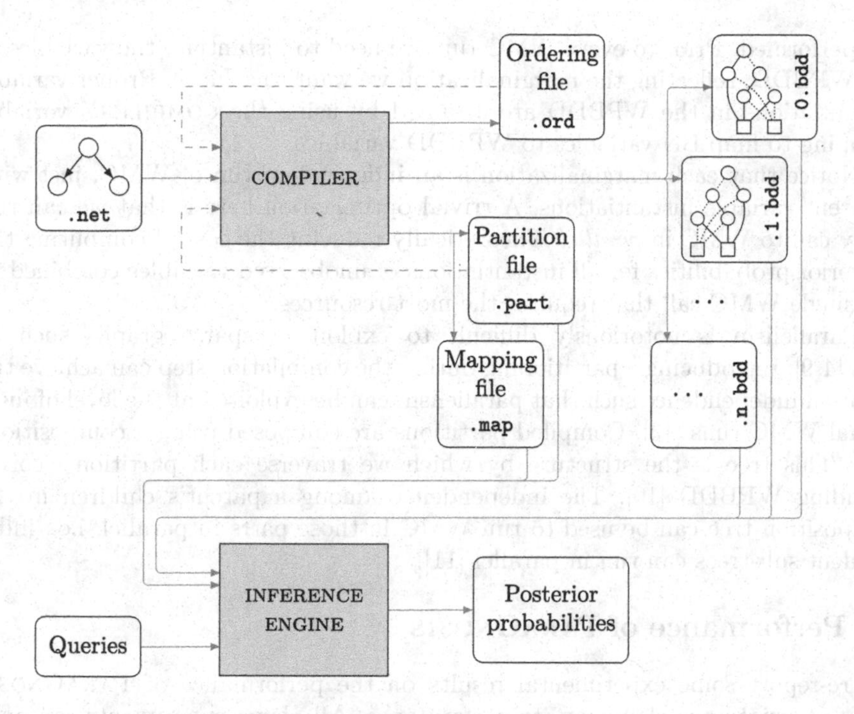

Fig. 6. PARAGNOSIS's architecture.

The COMPILER can introduce a partitioning to further improve overall performance [13]. With a user-provided number of partitions, a partitioning is found for the BN. Using simulated annealing, this partitioning is optimized by minimizing the sum of the tree-width of all partitions. Tree-width is a metric commonly used to indicate the complexity of BNs [6]. Optionally, a partitioning can also be provided by the user. With the partitioning in hand, the following steps can be performed in parallel, per partition. Theoretically, compilation is as fast as the slowest compiling partition [11].

Compilation is critically dependent upon a good variable ordering. The size of the resulting WPBDD (or any other ordered DD) is determined by it. The COMPILER can optionally be provided with a variable ordering as input. When not provided by one, an ordering is created automatically using the min-fill heuristic [5]. Further refinement can be attained when the user requires it. It is achieved by minimizing the tree-width of candidate orderings, using simulated annealing.

4.2 The INFERENCE ENGINE

The INFERENCE ENGINE takes as input probabilistic queries, and the COMPILER's output (Fig. 6). At its core, the INFERENCE ENGINE is able to perform probabilistic inference through marginalization. It does so using WMC (Sect. 3.2. Given a user-provided set of probabilistic queries, the corresponding marginalizations are performed. Prior to every WMC run, we need to instantiate the variables in the WPBDD, reflecting the marginalization we want to achieve. Proper variable instantiations in the WPBDD are attained by using the COMPILER's variable mapping to map BN variables to WPBDD variables.

Notice that each marginalization is an independent run of WMC, just with different variable instantiations. A trivial optimization here is that we can run every call to WMC in *parallel*. Theoretically reducing the cost of computing the posterior probabilities for all instantiations of unobserved variables combined to the single WMC call that requires the most resources.

Parallelism is notoriously difficult to exploit in sparse graphs such as BNs [4,9]. Introducing a partitioning during the compilation step can achieve the required independence such that parallelism can be exploited at the level of individual WMC runs [13]. Compiled partitions are composed using a composition-tree. This tree is the structure by which we traverse each partition's corresponding WPBDD [10]. The independence among a parent's children in the composition-tree can be used to run WMC in those parts in parallel, i.e., independent sub-trees can run in parallel [11].

5 Performance of ParaGnosis

We re-report some experimental results on the performance of ParaGnosis to substantiate our claims on its performance. All above experiments ran on a system with AMD Opteron 6376 processors, with 500+ Gb of RAM.

We have previously [10,12] compared WPBDD compilation costs to those of OBDDs and ZBDDs (using the CUDD 3.0.0 library), SDDs [14] (using the SDD 1.1.1 library), and OBDDs.[2,3,4] Much care is taken to create a fair comparison between libraries. We set up a head-to-head procedure, where each compiler was swapped-in and out. This ensures that the same steps are performed to produce each respective representation. The same ordering is thus tested with compiler and representation. Details of the experimental setup can be found in [10,13]. The results are again reported in Table 1 and show favorable runtimes for the WPBDD approach.

For inference, in previous work [10], we chose to compare to Ace (version 3.0) and to the junction tree algorithm using the publicly available Dlib library (version 18.18).[5,6] In previous works, ParaGnosis has also compared favorably to the HUGIN library (version 8.4).[7] [12]

We compared the aforementioned methods by measuring the speed by which they could solve the same set of probabilistic queries. Queries are created randomly, i.e., with a random number of observed variables and a random configuration. A query is created and fed to each method [10].

[2] CUDD is available at http://vlsi.colorado.edu/~fabio/.
[3] The SDD compiler is available at http://reasoning.cs.ucla.edu/sdd/.
[4] The WPBDD compiler is available at https://github.com/gisodal/paragnosis/.
[5] Ace is available at http://reasoning.cs.ucla.edu/ace/.
[6] Dlib is available at http://dlib.net/.
[7] HUGIN is available at http://www.hugin.com/.

Table 1. Compilation runtime (milliseconds), where $|\mathsf{at}(X)|$ are the number encoding variables for BN variables X, - implies compilation failure by exceeding 15 min or 500 Gb of RAM memory.

Bayesian Network	$\lvert\mathsf{at}(X)\rvert$	Partitioned t-WPBDD	t-WPBDD	OBDD	SDD	d-DNNF
sachs	24	0.148	**0.100**	1.932	29.119	92.179
student farm	25	0.117	**0.106**	1.403	4.646	118.641
printer ts	58	0.230	**0.198**	1.757	6.628	97.956
boblo	60	**0.213**	**0.213**	3.792	27.920	118.202
child	60	**0.195**	0.331	4.564	96.344	117.620
insurance	89	**0.494**	20.365	267.967	12337.980	680.771
weeduk	90	18.415	**6.091**	429.110	-	3472.012
alarm	105	**0.407**	0.467	10.085	400.158	157.163
water	116	**5.185**	1635.935	16034.149	-	1009.578
powerplant	120	**0.268**	0.361	9.409	119.856	159.193
carpo	122	0.426	**0.420**	13.910	119.122	137.955
win95pts	152	**0.874**	1.386	193.919	902.473	173.762
hepar2	162	**1.444**	1.567	414.316	31119.984	287.980
fungiuk	165	**22.186**	45.559	1667.940	-	12193.593
hailfinder	223	**1.061**	3.748	422.270	14350.353	354.151
3nt	228	**0.696**	2.397	344.902	4259.798	424.939
4sp	246	**0.849**	5.090	991.545	7041.476	573.015
barley	421	**611.830**	23290.743	-	-	-
mainuk	421	**584.409**	23443.483	-	-	-
andes	440	**3.267**	224.048	-	-	7785.916
pathfinder	520	**17.279**	18.057	22741.434	137591.643	2813.821
mildew	616	**42.611**	576.852	244920.444	-	885305.099
munin1	992	**11.929**	53899.548	-	-	-
pigs	1323	**4.444**	348.872	-	-	20623.511
link	1793	**174.897**	19412.863	-	-	-
diabetes	4682	**1297.221**	2622.924	-	-	-
munin2	5376	**96.809**	926.789	-	-	235544.805
munin3	5601	**52.350**	1088.710	-	-	102338.718
munin4	5645	**718.358**	2565.931	-	-	162054.255
munin	5651	**1407.531**	2360.196	-	-	161133.160

The results are again reported in Table 2 and show favorable runtimes for the WPBDD approach, even for the partitioned approach. It shows the average runtime of a query across the aforementioned set of tested queries. While partitioning in theory can shift some workload from compilation to inference, it can also further reduce the size of the knowledge base, which explains these results.

In future experimental work, it would be interesting to compare ParaGnosis against Mora [2] which takes the new approach by reducing the inference problem to a probabilistic program. This new approach has not yet been compared to any of the above tools.

Table 2. Inference runtime averaged per query (milliseconds), where - implies inference failure by exceeding 15 min, or compilation failure.

Bayesian Network	Partitioned t-WPBDD	t-WPBDD	d-DNNF	DLIB
sachs	0.023	**0.011**	2.975	?
student farm	0.035	**0.016**	2.813	2627.214
printer ts	0.007	**0.006**	2.852	?
boblo	0.054	**0.034**	3.713	3545.921
child	0.345	**0.036**	5.695	3545.921
insurance	7.486	**1.874**	36.884	?
weeduk	0.607	**0.262**	30.908	3586.316
alarm	0.187	**0.115**	6.513	3547.504
water	**25.176**	74.135	33.512	-
powerplant	**0.025**	0.032	6.249	3393.056
carpo	0.138	**0.037**	5.739	3515.298
win95pts	**0.371**	0.635	9.680	3220.060
hepar2	**0.247**	1.133	18.659	3463.509
fungiuk	18.822	**5.290**	42.814	3567.136
hailfinder	25.137	**1.747**	19.618	2448.653
3nt	18.411	**8.250**	21.559	?
4sp	5.748	**1.277**	30.043	?
barley	**1399.542**	1798.278	-	-
mainuk	**1377.117**	1782.512	-	-
andes	178.610	185.205	**144.691**	?
pathfinder	5.394	**0.639**	30.686	?
mildew	351.788	552.496	**208.582**	-
munin1	7183.857	**6836.045**	-	-
pigs	248.866	**70.266**	179.088	-
link	**9893.431**	-	-	-
diabetes	968.839	**618.687**	-	-
munin2	**107.788**	207.768	384.055	-
munin3	828.535	**140.398**	263.751	-
munin4	**280.275**	318.687	402.651	-
munin	377.117	**302.675**	416.733	-

6 Discussion

ParaGnosis shows that parallelism can benefit knowledge compilation and inference for Bayesian networks. Our chosen parallelization approach is based on partitioning. This approach has previously been used to speed up symbolic model checking [19,25,28]. We additionally find that the partitioning introduces a tradeoff between compilation times and inference times, sacrificing some performance in inference to gain parallel scalability.

The computational complexity of inference is linear in the size of the target representation [16]. However, it increases when partitioning is employed. A compiled partition in a composition-tree must be exponentially traversed in the size of the cutset that separates it from its parent. One traversal for each instantiation of the cutset variables. However, this is compensated by a number of principles. (1) the combined size of all partition BDDs is significantly reduced compared to the monolithically compiled BDD [13]; (2) partition BDDs only represent a portion of its total, reducing traversal resources; (3) each child instantiation can be traversed in parallel, potentially reducing traversal cost of exponential traversals to the cost of 1 [10]; (4) as partition BDDs are small, cache locality start to play an important role, giving an advantage over monolithic BDDs [11]. Beyond the complexity introduced by cutsets, inference remains linear. Small cutsets can therefore play a crucial role in performance. Separate empirical investigations on partitioning and parallelism show their respective contribution to ParaGnosis's performance [10,11].

The parallelization approach of ParaGnosis is orthogonal to the chosen target representation of the knowledge base, which we demonstrated by using it for four different target representations [11]. As a consequence, the approach is to a certain extent orthogonal to the exploitation of local structure [30] by those representations, as local structure can still be exploited within the partitioned subproblems. For instance, we showed that causal dependence is fully exploited when using decision diagram representations in the partitions.

For target representations, many choices exist [5,7,14]. Since our partitioning technique exploits the treewidth [6] of the representation [10, §5], and the algorithms are based on message passing [10, §4], other representations like ADD and d-DNNF can be parallelized alike. While our earlier work [11] compared the performance of ParaGnosis against various of these other representations, showing competitive performance, here we will point out some differences between the other representations and suggest future work.

Like AADD [29], and its similar cousins SLDD [34], FEVBDD [32] and QMDD [24], our target representation WPBDD factors out probabilities on the edges of the diagram, resulting in more succinctness than for instance achieved with ADD [1], QuiDD [33] and MTDD [8]. However, unlike AADD, it only factors according to the structure of the Bayesian network, which sacrifices succinctness but ensures exact representation of the floating point numbers. The latter can be quite important in practice, as rounding errors from manipulation operations can rapidly propagate in the discrete data structure, resulting in numerical instability [21,27,35].

The effects of different variable orders are known to be crucial in many representation languages. Representations like d-DNNF [5] and SDD [14], allow more freedom in the order and could potentially improve the results of ParaGnosis.

Early versions of ParaGnosis also tried different parallelization approaches, like the fine-grained task-based scheduling of Sylvan [18], which has shown that good parallel scalability is possible for model checking problems in BDDs, ADDs (also called Multi-Terminal DDs), and MDDs (Sylvan uses a version called List DD [19]). In future work, we hope to establish why this approach did not yield good performance for knowledge compilation as well.

References

1. Bahar, R.I., et al.: Algebraic decision diagrams and their applications. In: Proceedings of 1993 International Conference on Computer Aided Design, pp. 188–191 (1993)
2. Bartocci, E., Kovács, L., Stankovič, M.: Mora - automatic generation of moment-based invariants. In: TACAS 2020. LNCS, vol. 12078, pp. 492–498. Springer, Cham (2020). https://doi.org/10.1007/978-3-030-45190-5_28
3. Bryant, R.E.: Graph-based algorithms for Boolean function manipulation. Trans. Comput. **100**, 677–691 (1986)
4. Ceylan, I.I., Darwiche, A., Van den Broeck, G.: Open-world probabilistic databases: semantics, algorithms, complexity. Artif. Intell. **295**, 103–123 (2021)
5. Chavira, M., Darwiche, A.: On probabilistic inference by weighted model counting. Artif. Intell. **172**, 772–799 (2008)
6. Chen, Y., Darwiche, A.: On the definition and computation of causal treewidth. In: The 38th Conference on Uncertainty in Artificial Intelligence (2022)
7. Choi, Y.K., Santillana, C., Shen, Y., Darwiche, A., Cong, J.: FPGA acceleration of Probabilistic Sentential Decision Diagrams with high-level synthesis. ACM Trans. Reconfigurable Technol. Syst. **16**(2), 1–22 (2022)
8. Clarke, E.M., Fujita, M., McGeer, P.C., McMillan, K., Yang, J.C.-Y., Zhao, X.: Multi-terminal binary decision diagrams: An efficient data structure for matrix representation (2001)
9. Dal, G.H., Kosters, W.A., Takes, F.W.: Fast diameter computation of large sparse graphs using GPUs. In: International Conference on Parallel, Distributed and Network-Based Processing, pp. 632–639 (2014)
10. Dal, G.H., Laarman, A.W., Hommersom, A., Lucas, P.J.: A compositional approach to probabilistic knowledge compilation. Int. J. Approximate Reason. **138**, 38–66 (2021)
11. Dal, G.H., Laarman, A.W., Lucas, P.J.F.: Parallel probabilistic inference by weighted model counting. In: International Conference on Probabilistic Graphical Models, pp. 97–108 (2018)
12. Dal, G.H., Lucas, P.J.F.: Weighted positive binary decision diagrams for exact probabilistic inference. Int. J. Approx. Reason. **90**, 411–432 (2017)
13. Dal, G.H., Michels, S., Lucas, P.J.F.: Reducing the cost of probabilistic knowledge compilation. Mach. Learn. Res. **73** 141–152 (2017)
14. Darwiche, A.: SDD: A new canonical representation of propositional knowledge bases. In: International Joint Conference on Artificial Intelligence, vol. 22, pp. 819 (2011)
15. Darwiche, A., Hirth, A.: On the (complete) reasons behind decisions. J. Logic, Lang. Inform. Ages **18**, 1–26 (2022)
16. Darwiche, A., Marquis, P.: A knowledge compilation map. J. Artif. Intell. Res. **17**, 229–264 (2002)

17. Darwiche, A., Marquis, P.: On quantifying literals in Boolean logic and its applications to explainable AI. J. Artif. Intell. Res. **72**, 285–328 (2021)
18. van Dijk, T., Laarman, A., van de Pol, J.: Multi-core BDD operations for symbolic reachability. Electron. Notes Theor. Comput. Sci. **296**, 127–143 (2013)
19. van Dijk, T., van de Pol, J.: Sylvan: multi-core framework for decision diagrams. Int. J. Softw. Tools Technol. Transfer **19**, 675–696 (2017)
20. Dudek, J., Phan, V., Vardi, M.: ADDMC: Weighted model counting with algebraic decision diagrams. In: International Conference on Artificial Intelligence, pp. 1468–1476 (2020)
21. Hillmich, S., Burgholzer, L., Stogmuller, F., Wille, R.: Reordering decision diagrams for quantum computing is harder than you might think. In: International Conference on Reversible Computation, pp. 93–107 (2022)
22. Hitzler, P., Sarker, M.K.: Tractable Boolean and arithmetic circuits. Neuro-Symbolic Artif. Intell.: State Art **342**, 146 (2022)
23. Madsen, A.L., Lang, M., Kjaerulff, U.B., Jensen, F.: The Hugin tool for learning Bayesian networks. In: European Conference on Symbolic and Quantitative Approaches to Reasoning and Uncertainty, pp. 594–605 (2003)
24. Miller, D.M., Thornton, M.A.: QMDD: A decision diagram structure for reversible and quantum circuits. In: 36th International Symposium on Multiple-Valued Logic, pp. 30–36 (2006)
25. Narayan, A., Jain, J., Fujita, M., Sangiovanni-Vincentelli, A.: Partitioned ROBDDs-a compact, canonical and efficiently manipulable representation for Boolean functions. In: Proceedings of International Conference on Computer Aided Design, pp. 547–554 (1996)
26. Judea Pearl. Bayesian networks (2011)
27. Peham, T., Burgholzer, L., Wille, R.: Equivalence checking paradigms in quantum circuit design: A case study. In: Proceedings of the 59th Design Automation Conference, pp. 517–522 (2022)
28. Sahoo, D.: A partitioning methodology for BDD-based verification. In: International Conference on Formal Methods in Computer-Aided Design, pp. 399–413 (2004)
29. Sanner, S., McAllester, D.: Affine algebraic decision diagrams (AADDs) and their application to structured probabilistic inference. In: International Joint Conference on Artificial Intelligence 2005, pp. 1384–1390 (2005)
30. Shih, A., Choi, A., Darwiche, A.: Compiling Bayesian network classifiers into decision graphs. In: International Conference on Artificial Intelligence, vol. 33, pp. 7966–7974 (2019)
31. Sokukcu, M., Sakar, C.: Risk analysis of collision accidents during underway sts berthing maneuver through integrating fault tree analysis (FTA) into Bayesian network. Appl. Ocean Res. **126**, 103–124 (2022)
32. Tafertshofer, P., Pedram, M.: Factored edge-valued binary decision diagrams. Formal Methods Syst. Design **10**(2), 243–270 (1997)
33. Viamontes, G.F., Markov, I.L., Hayes, J.P.: Improving gate-level simulation of quantum circuits. Quant. Inform. Process., **2**, 347–380 (2003)
34. Wilson, N.: Decision diagrams for the computation of semiring valuations. In: Proceedings of the 19th International Joint Conference on Artificial Intelligence, pp. 331–336 (2005)
35. Zulehner, A., Hillmich, S., Wille, R.: How to efficiently handle complex values? Implementing decision diagrams for quantum computing. In: 2019 International Conference on Computer-Aided Design, pp. 1–7 (2019)

Concurrency

Model Checking Futexes

Hugues Evrard[1(✉)] and Alastair F. Donaldson[2]

[1] Google, Paris, France
hevrard@google.com
[2] Imperial College London, London, UK
alastair.donaldson@imperial.ac.uk

Abstract. The *futex* Linux system call enables implementing performant inter-thread synchronisation primitives, such as mutexes and condition variables. However, the futex system call is notoriously tricky to use correctly. In this case study, we use the Spin model checker to verify safety properties of a number of futex-based mutex and condition variable implementations. We show how model checking is able to detect bugs that affected real-world implementations, and confirm current implementations are correct. The Promela models we have developed are available as open source, and may be useful as teaching material for classes that cover futex-based synchronisation primitives, and as a template on how to perform formal verification on new synchronisation primitive designs.

Keywords: futex · mutual exclusion · condition variables · model checking · Promela/Spin

1 Introduction

The futex system call was introduced to the Linux kernel in the early 2000 s s in order to support efficient synchronisation primitives [9]. The name "futex" is derived from "**f**ast **u**serspace mu**tex**", because one of the most important use cases for the futex system call is the efficient implementation of mutexes, striking a balance between OS semaphores, whose manipulation always involves a system call even when contention is low, and spinlocks, which operate entirely in userspace but may lead to high CPU usage when contention is high.

When used in a careful and clever manner, futexes can enable efficient inter-thread and inter-process synchronisation. However, futexes are also notoriously tricky to use correctly. According to Drepper, in his aptly-titled paper "Futexes are Tricky" [7], a package authored by one of the inventors of the futex system call, containing user-level code demonstrating its use, turned out to be incorrect. Drepper describes why an early mutex implementation suffers from correctness problems, and presents two alternative implementations, arguing their correctness informally. In an article on futex-based condition variables [6], Denis-Courmont describes a number of flawed proposals for implementing condition variables, and a proposal that is argued to be correct under reasonable practical assumptions.

A limitation of these expositions of futex-based synchronisation primitives is that they are based on informal descriptions of how code snippets might behave in a concurrent context. The reader may not fully understand the (often subtle) arguments for

G. Caltais and C. Schilling (Eds.): SPIN 2023, LNCS 13872, pp. 41–58, 2023.
https://doi.org/10.1007/978-3-031-32157-3_3

(in)correctness, and even if they do, it may be hard for them to imagine the consequences of alternative implementation choices.

In this case study, we investigate the use of the Promela language and Spin model checker [13] to express and analyse various proposals from [7] and [6] for futex-based mutexes and condition variables, respectively. Due to the ability of model checking to produce counterexamples, our Promela models of incorrect implementations lead to step-by-step traces that illustrate bug-triggering thread interleavings. This facility also aids in understanding why certain details of correct implementations are important, because one can change those details and inspect the counterexamples that arise as a result. In particular, we show that model checking can detect bugs that affected real-world implementations of mutexes and that it can confirm bugs in both naive and real-world implementations of condition variables. We also show that model checking aids in understanding the importance of certain intricacies of a futex-based mutex design.

The Promela models we have developed are available as open source, together with instructions on how to use Spin to analyse them [8]. We envisage that they may be useful as teaching material in classes that cover futex-based synchronisation primitives. In fact, our investigation into the application of model checking to this problem was inspired by the experience of one of the authors teaching about futex-based mutexes on a course at Imperial College London, and being dissatisfied with his informal correctness-related explanations. We also hope that our models will serve as a template on how to perform formal verification on new synchronisation primitive designs.

The rest of the paper is organised as follows. In Sect. 2 we provide necessary background on the futex system call. We explain how we have modelled this system call in Promela, to enable the modelling of synchronisation primitives that use it, in Sect. 3. Our Promela models of mutexes and condition variables rely on the modelling of various integer atomic operations, including operations that may overflow; we discuss these in Sect. 4. In Sect. 5 we work through examples of futex-based mutex implementations from Drepper's paper [7], explaining how we have modelled each mutex variant using Promela and presenting insights into our analysis of these models using Spin. In Sect. 6 we turn to condition variables, working through some implementation proposals from Denis-Courmont's article [6]. We discuss related work in Sect. 7 and conclude with a discussion of future directions in Sect. 8.

Throughout the paper we assume the reader is familiar with Promela and with basic operation of the Spin model checker. See [13] for a definitive reference.

2 The Futex System Call

The word *futex* is often used to designate three things: (1) a 32-bit addressable value also called the *futex word*, (2) the futex system call, and (3) mutex implementations based on the futex system call. In this section, we are concerned with (1) and (2), while (3) is discussed in Sect. 5.

Generally speaking, the futex system call enables threads to block depending on the value of a given memory word—the futex word—or to wake up threads that are waiting in a queue associated with a futex word. In practice, a futex system call has the following form:[1]

```
long syscall(SYS_futex,
  uint32_t *addr, // pointer to the futex word
  int futex_op,   // operation: FUTEX_WAIT, FUTEX_WAKE, ...
  uint32_t val,   // plain value argument
  ...); // extra arguments for other operations
```

It is multiplexed via its `futex_op` argument, which refers to one of various operations. In this case study, we focus on the two basic operations: `FUTEX_WAIT` and `FUTEX_WAKE`, where only the `addr` and `val` arguments are relevant.

`FUTEX_WAIT`: the calling thread blocks and goes to sleep only if the value of the futex word addressed by `addr` is equal to the plain value argument `val`. This operation is atomic with respect to the futex word, which is typically in memory shared between threads. Similar to compare-and-exchange instructions on atomics, this call has a compare-and-block semantics: loading the futex word's value, comparing it to `val`, and blocking happen atomically and are totally ordered with respect to other concurrent operations on the futex word.

`FUTEX_WAKE`: the calling thread wakes threads waiting on the futex identified by `addr`. It wakes `val` threads, or the amount of threads waiting on `addr`, whichever is smaller. There is no guarantee on which threads are woken up, or in which order threads are woken up.

The name "futex" is derived from **f**ast and **u**serspace because futex-based synchronisation primitive implementations (such as implementations of mu**tex**es) typically try first to synchronise using userspace atomic operations on a shared futex word, and only resort to futex system calls in case of contention. We see this pattern in the mutex implementations in Sect. 5.

3 Modelling the Futex System Call Variants

We model futexes in Promela as a `Futex` type, and two inline macros `futex_wait` and `futex_wake` to represent these variants of the general system call. Before covering these in detail, some general remarks about our modelling approach. To keep the state vector size under control, we use `byte` values virtually everywhere we would use `int` values in C: this is without loss of generality since, in our examples with a handful of threads, all interesting values are within [0, 255]. Threads are mapped to Promela's `proctype` and are identified by their `_pid` builtin variable. The total number of threads is a global constant that we use to dimension arrays. It is defined by a preprocessor macro, `NUM_THREADS`, so that it can be easily changed when invoking Spin (e.g. `spin -DNUM_THREADS=5 ...`):

```
#ifndef NUM_THREADS
#define NUM_THREADS 2
#endif // NUM_THREADS
```

Now, on to futexes: the `Futex` type contains a futex word, the list of threads that are waiting on this futex, and a counter of currently waiting threads:

[1] https://man7.org/linux/man-pages/man2/futex.2.html.

```
1 typedef Futex {
2   byte word; // Futex word
3   bool wait[NUM_THREADS]; // Wait list: array of bool indexed by thread IDs,
4                           // thread T is waiting iff wait[T] is true
5   byte num_waiting; // Number of threads currently waiting
6 }
```

The wait list is modelled via an array indexed by thread IDs: this will prove convenient to wake up sleeping threads in a non-deterministic order. In a C program, each futex is identified by the address of its futex word; here each futex is identified by a variable of type `Futex` which is in global scope so that all threads can refer to it.

The `futex_wait` inline macro models the `FUTEX_WAIT` operation:

```
1 inline futex_wait(futex, val) {
2   if
3   :: d_step {
4        futex.word == val ->
5        printf("T%d futex_wait, value match: %d; sleep\n",
6               _pid, futex.word);
7        assert(!futex.wait[_pid]); // The thread must not be sleeping already
8        futex.wait[_pid] = true;
9        futex.num_waiting++;
10     }
11     d_step { !futex.wait[_pid] -> printf("T%d has woken\n", _pid); }
12  :: d_step {
13       else -> printf("T%d futex_wait, value mismatch: %d vs. %d; do not sleep\n
     ",
14                      _pid, futex.word, val);
15     }
16   fi
17 }
```

It takes as argument a variable of type `Futex`, and a plain value to compare to the futex word. If they are equal, the thread goes to sleep: we set its entry in the wait list, and increment the counter of waiting threads. An assertion checks that only non-sleeping threads may go to sleep. Then, the thread blocks until its wait list entry is set to false. If the value argument differs from the futex word, then the thread continues without blocking. Log messages prefixed by the ID of the executing thread are printed to ease the understanding of counterexamples.

The atomic compare-and-block semantics is achieved with the first `d_step` (line 3): this is a better choice than `atomic`, since all the statements in a `d_step` are treated as a single state change by Spin, thus reducing the search depth. It is safe to use `d_step` over `atomic` here since all contained statements are deterministic, there is no jump in or out the `d_step` scope, and there is no blocking statement in the middle of the scope.

The `d_step` blocks at lines 11 and 12 guarantee that logging prints values related to the state in which a thread is woken up, or in which a value mismatch occurs, respectively (to avoid confusion due to log messages from other threads being interleaved).

The `futex_wake` inline macro models the `FUTEX_WAKE` operation:

```
1 inline futex_wake(futex, num_to_wake) {
2   atomic {
3     assert(!futex.wait[_pid]); // The waker must not be asleep
4     byte num_woken = 0;
5     do
6     :: num_woken == num_to_wake || futex.num_waiting == 0 ->
```

```
7         break
8     :: else ->
9         if
10        :: futex.wait[0] -> futex.wait[0] = false; printf("T%d wakes T0\n", _pid)
11        :: futex.wait[1] -> futex.wait[1] = false; printf("T%d wakes T1\n", _pid)
12 #if NUM_THREADS > 2
13        :: futex.wait[2] -> futex.wait[2] = false; printf("T%d wakes T2\n", _pid)
14 #endif
15 #if NUM_THREADS > 3
16        :: futex.wait[3] -> futex.wait[3] = false; printf("T%d wakes T3\n", _pid)
17 #endif
18 #if NUM_THREADS > 4
19        :: futex.wait[4] -> futex.wait[4] = false; printf("T%d wakes T4\n", _pid)
20 #endif
21 #if NUM_THREADS > 5
22 #error "NUM_THREADS > 5, add more if branches in futex_wake"
23 #endif
24        fi
25        futex.num_waiting--;
26        num_woken++;
27    od
28    printf("T%d woke up %d thread(s)\n", _pid, num_woken);
29    num_woken = 0; // Reset to avoid state space explosion
30  }
31 }
```

The num_to_wake argument indicates the number of threads to wake up, the local variable num_woken counts how many threads have been woken so far. Note that we cannot eliminate num_woken and instead decrement num_to_wake until it reaches zero since the macro argument num_to_wake may be a literal value, e.g. in a call such as futex_wake(futex, 1). We enter a loop that wakes one thread per iteration, until the desired number of threads have been woken or there are no more threads to wake. When waking a thread, we use a nondeterministic if to pick one of the sleeping threads, which is then woken up by setting its entry in the futex wait list array to false.

The whole macro body is contained in an atomic scope to prevent concurrent accesses to the futex internals. This time, d_step cannot be used due to the non-deterministic order in which threads are woken. At the end of the atomic scope, num_woken is reset to zero. This is vital to reduce state-space explosion: it prevents Spin from regarding otherwise identical states that differ only in the final value of num_woken as distinct, which would lead to Spin continuing its exhaustive search from each such state.

Relying on the non-deterministic selection of enabled if branches requires exactly NUM_THREADS branches: we use the C preprocessor to achieve this, supporting here up to five threads, with it being easy to support more threads by adding further if branches. For a really arbitrary number of threads, one could easily script the generation of these branches. We opt for the C preprocessor to keep the Promela code self-contained.

4 Modelling Atomic Operations and Overflow

The mutex and condition variable implementations rely on standard C/C++ atomic operations that we model in Promela. Atomic compare-and-exchange, cmpxchg, compares the value at a location with an expected value: if they match, the location is set to a desired value; otherwise it is left unchanged. Either way, the original location value is returned, here via a result parameter:

```
inline cmpxchg(location, expected, desired, result) { d_step {
  result = location; location = (location == expected -> desired : location)
}}
```

Atomic fetch-and-increment, `fetch_inc`, returns the current value of a location before incrementing it. To limit both state space explosion and counterexample length, we model overflow and wrapping on `byte` values with a tighter upper bound set to the total number of threads plus one, represented a constant, `MAX_BYTE_VALUE`. This is without loss of generality, since C/C++ atomic integers also wrap upon overflow. We define the `inc` macro to handle overflow, and use `d_step` to make `fetch_inc` atomic:

```
#define MAX_BYTE_VALUE (NUM_THREADS + 1)
#define inc(a)   (a == MAX_BYTE_VALUE -> 0 : a + 1)
inline fetch_inc(location, result) {
  d_step { result = location; location = inc(location) }
}
```

In a similar fashion, we define a `dec` macro that handles underflow, and a `fetch_dec` macro for atomic fetch-and-decrement. Some of the Promela models discussed later also make direct use of the `inc` macro when performing an increment in a local expression, rather than operating on a futex word.

5 Model Checking Futex-based Mutexes

We describe the usage scenario and properties for mutexes to which model checking is applied (Sect. 5.1), then the modelling and verification of the two main mutex implementations from [7] (Sects. 5.2 and 5.3).

5.1 Model Checking Harness and Properties

We use the following harness to enable model checking of various futex-based mutex implementations:

```
byte num_threads_in_cs; // Number of threads in the critical section (CS)

active [NUM_THREADS] proctype Thread() {
  do
  :: lock();
     num_threads_in_cs++;
     num_threads_in_cs--;
     unlock();
  :: printf("T%d is done\n", _pid) -> break
  od
}

ltl safe_cs { [](num_threads_in_cs <= 1) } // Never more than one thread in CS
```

It uses an `active` proctype to launch `NUM_THREADS` threads, each of which uses the `lock()` and `unlock()` inline macros to repeatedly lock and unlock a shared mutex. Separate versions of these macros are provided for each mutex implementation discussed below. The macros assume that a global variable of type `Futex` is available.

Global variable num_threads_in_cs, initialised to 0 by default, is used to record when threads enter and leave the critical section.

We consider model checking of two safety properties: (1) freedom from invalid end states (a built-in feature of Spin), which confirms that it is not possible for a thread to become blocked in a call to futex_wait when all other threads have terminated, and (2) mutual exclusion, captured by the "safe critical section" linear temporal logic (LTL) property, safe_cs, which checks that the number of threads in the critical section never exceeds one.

5.2 Incorrect Futex-based Mutex

The following shows C++ code for a subtly incorrect futex-based mutex, adapted from [7, §4]. The futex word is the 32-bit atomic integer field futex_word. The intention is that the mutex is free if and only if futex_word has value 0.

```
1  class Mutex {
2  public:
3    Mutex() : futex_word(0) {}
4    void lock() {
5      uint32_t old_value;
6      while ((old_value = futex_word.fetch_add(1)) != 0)
7        futex_wait(&futex_word, old_value + 1);
8    }
9    void unlock() {
10     futex_word.store(0);
11     futex_wake(&futex_word, 1);
12   }
13
14 private:
15   atomic<uint32_t> futex_word;
16 };
```

A thread attempts to lock the mutex by incrementing futex_word via a fetch_add, storing the previous value of the futex word in the local variable old_value. If this value is 0 then the thread has locked the mutex, by changing futex_word from 0 to 1, and can return from lock. Otherwise, the thread calls futex_wait with old_value + 1: if no other thread modifies the futex word in between the call to fetch_add and the call to futex_wait, this value will match the futex word and the thread will go to sleep until the lock becomes free. If another thread modifies the futex word before the call to futex_wait, then this call will not put the first thread to sleep so that the thread will immediately attempt to acquire the mutex again via another fetch_add.

Unlocking the mutex is simpler: futex_word is set to 0, and futex_wake is called so that one of the threads waiting on futex_word, if any, will be woken.

Drepper discusses a correctness issue triggered by an overflow of the futex word.

Suppose several threads are contending to try to lock an already locked mutex. It is possible that while a given contending thread T1 is between the calls to fetch_add and futex_wait, another contending thread T2 calls fetch_add and modifies the futex word, such that T1 will not go to sleep and will itself call fetch_add again, preventing T2 from going to sleep. This can go on until the futex word wraps back to 0, in which case a contending thread might believe it can successfully lock the mutex.

This mutex design is modelled in Promela by the following inline macros:

```
inline lock() {
  byte old_value;
  do
  :: atomic {
       fetch_inc(futex.word, old_value);
       if
       :: old_value == 0 -> printf("T%d locks mutex\n", _pid); break
       :: else -> printf("T%d lock fail, old_value: %d\n", _pid, old_value);
       fi
     }
     futex_wait(futex, inc(old_value))
  od
}

inline unlock() {
  d_step { futex.word = 0; printf("T%d unlocks mutex\n", _pid); }
  futex_wake(futex, 1);
}
```

Here, we make use of atomic and d_step blocks to (a) ensure that print statements are executed atomically with the actions that they aim to document, and (b) limit state explosion by allowing interleavings only between operations that have inter-thread visibility: calls to futex_wait/futex_wake, and statements that manipulate the futex word. For example, it is vital that there is an interleaving point between fetch_inc at line 5 and futex_wait at line 11. However, there is no value in considering thread interleavings between the fetch_inc and the if..fi that immediately follows. These only involve a thread manipulating its local state. An interleaving point will cause needless state-space explosion which we have found Spin's partial order reduction does not completely alleviate.

With two threads, Spin quickly verifies the safe_cs property and confirms that all end states are valid. This is expected: the bug described above requires a race between multiple contending threads when the mutex is already held by a further thread. With three threads, Spin quickly reports a counterexample (minimised using Spin's iterative shortening algorithm) with the following messages:

```
T0 locks mutex
T1 lock fail, old_value: 1
T2 lock fail, old_value: 2
T1 futex_wait, value mismatch: 3 vs. 2; do not sleep
T1 lock fail, old_value: 3
T2 futex_wait, value mismatch: 4 vs. 3; do not sleep
T2 lock fail, old_value: 4
T1 futex_wait, value mismatch: 0 vs. 4; do not sleep
T1 locks mutex
   assertion num_threads_in_cs <= 1 violated
```

This nicely illustrates the problem where threads T1 and T2 repeatedly prevent one another from sleeping by each incrementing the futex word before the other can call futex_wake; "value mismatch: 0 vs. 4" shows the futex word wrapping from 4 to 0.

The "no invalid end states" property also fails, though with a longer counterexample. Here is a summary of the problem. Suppose that T0 holds the lock. T1 and T2 then get into a race, incrementing the futex word until T1 observed the word's old value to be 3 and T2 observed the word's old value to be 4, so that the word's *current* value is 0 (T2 having caused it to wrap-around). T1 is poised to call `futex_wait(4)`, and T2 is poised to call `futex_wait(0)`, but neither have done so yet.

At this point, T0 unlocks the mutex by setting the futex word to 0, wakes up no threads, and terminates. T1 calls `futex_wait(4)`, which immediately returns due to a value mismatch; T1 tries and succeeds to lock the mutex, then immediately releases it, waking up no threads, and terminates. T2 finally calls `futex_wait(0)`, and by now the futex word value *is* 0, so T2 goes to sleep with no chance of being woken since all other threads have terminated.

As explained in [7], this problem affected real code. It is great that model checking can quickly expose it, with a clear counterexample.

5.3 Correct Futex-based Mutex

Drepper goes on to present the following more intricate mutex implementation compared with that of Sect. 5.2, which is claimed to be correct [7, §5]:

```
1  class Mutex {
2  public:
3    Mutex() : futex_word(0) {}
4    void lock() {
5      uint32_t old_value;
6      if ((old_value = cmpxchg(futex_word, 0, 1)) != 0)
7        do {
8          if (old_value == 2 || cmpxchg(futex_word, 1, 2) != 0)
9            futex_wait(&futex_word, 2);
10       } while ((old_value = cmpxchg(futex_word, 0, 2)) != 0);
11   }
12   void unlock() {
13     if (futex_word.fetch_sub(1) != 1) {
14       futex_word.store(0);
15       futex_wake(&futex_word, 1);
16     }
17   }
18
19 private:
20   atomic<uint32_t> futex_word;
21 };
```

We use *waiters* to refer to threads that are asleep due to having called `futex_wait`. In this implementation, the futex word can take on one of three values. A value of 0 means that the mutex is free, while values 1 and 2 mean that some thread, say T, holds the mutex. If the futex word is 1, a state referred as "locked, no waiters", then when T unlocks the mutex, T is not obliged to wake up any waiters. In contrast, if the futex word is 2, a state referred as "locked, waiters", then when T unlocks the mutex, T must call `futex_wake` to request that one waiter be woken.

In `lock`, a thread T first tries to lock the mutex by changing the value of the futex word from 0 to 1 via a `cmpxchg` at line 6. If T succeeds in doing this then it has locked the mutex and can return. In this case, we say that the thread has locked the mutex on the *fast path*.

Otherwise, T must contend for the mutex on the *slow path*, via the loop headed at line 7. The thread considers calling `futex_wait` to go to sleep and be notified when the mutex becomes free. Before this, at line 8, T checks whether the previous value of the futex word was already 2 ("locked, waiters"). If not, the previous value must have been 1 (“locked, no waiters”), so T attempts to change the value from 1 to 2 via another `cmpxchg`. Normally T will then call `futex_wait` at line line 9, but if the `cmpxchg` returns a previous value of 0 this indicates that the mutex has suddenly become free, in which case there is no point calling `futex_wait`; instead, T should try again to lock the mutex.

Once T returns from `futex_wait`, or if T decided not to perform this call due to observing the mutex to be free, it performs another `cmpxchg` to try to lock the mutex at line 10. In contrast to line 6, here T attempts to change the futex word from 0 to 2 to record the fact that T had to contend for the mutex and so there may be some waiters. T leaves the loop only when the `cmpxchg` at line 10 returns 0: we say that T has locked the mutex *on the slow path*.

The `unlock` function is simpler: the futex word is atomically decremented and its old value is inspected (line 13). If the old value is 1, “locked, no waiters”, then the futex word is now 0 so the mutex is properly unlocked, and the thread has no obligation to wake up waiters, so can return from `unlock`. Otherwise the old value must have been 2, “locked, waiters”, so the thread must set the futex word to 0 (line 14) and call `futex_wake` to wake up one waiter, if any (line 15).

This mutex design is difficult to understand, and it is unlikely that a reader will gain a full understanding from a best-effort prose explanation such as the above, or the explanation given by Drepper [7]. Particularly subtle is the fact that the futex word can have value 1, “locked, no waiters”, despite the fact that there *are* waiters, and conversely the mutex word can have value 2, “locked, waiters” even though there are *no* waiters. Reasoning informally that this mutex implementation is correct is difficult, hence why we decided to model it formally. Here are the Promela `lock` and `unlock` macros for this mutex implementation:

```
1  inline lock() {
2    byte old_value;
3    atomic {
4      cmpxchg(futex.word, 0, 1, old_value);
5      if
6      :: old_value == 0 -> printf("T%d locks mutex on fast path\n", _pid);
7         goto acquired_mutex
8      :: else -> printf("T%d fails to lock mutex on fast path\n", _pid)
9      fi
10   }
11   do
12   :: atomic {
13        if
14        :: old_value == 2
15        :: else -> assert(old_value == 1);
16           cmpxchg(futex.word, 1, 2, old_value)
17           if
18           :: old_value == 0 -> goto retry
19           :: else
20           fi
21        fi
22      }
23      futex_wait(futex, 2)
24 retry:
25      atomic {
26        cmpxchg(futex.word, 0, 2, old_value)
27        if
28        :: old_value == 0 -> printf("T%d locks mutex on slow path\n", _pid);
29           goto acquired_mutex
30        :: else -> printf("T%d fails to lock mutex on slow path\n", _pid)
```

```
31      fi
32        }
33    od
34    acquired_mutex:
35  }
36
37  inline unlock() {
38    byte old_value;
39    d_step {
40      fetch_dec(futex.word, old_value);
41      printf("T%d decrements futex word from %d to %d\n", _pid, old_value, futex.
         word);
42    }
43    if
44    :: d_step { old_value == 2 -> futex.word = 0; old_value = 0 }
45       futex_wake(futex, 1)
46    :: d_step { old_value == 1 -> old_value = 0 }
47    fi
48  }
```

As with the Promela code of Sect. 5.2 we use print statements for counterexample readability and use `atomic` and `d_step` so that threads only interleave after issuing visible operations. The Promela is a fairly straightforward reflection of the original C++, but the differences in the structured control flow constructs offered by the language led to us making use of Promela's `goto`.

Checking Correctness. Spin is able to rapidly verify the `safe_cs` property, as well as freedom from invalid end states (see Sect. 5.1) for our model of this mutex implementation for up to five threads. The results for checking `safe_cs` are summarised in Table 1, checking invalid end states leads to the same number of states and similar times, so they are omitted. Results were obtained using Spin version 6.5.2 on an AMD EPYC workstation running Linux 5.19, with C code generated by Spin compiled using GCC 12.2.0. The times shown are averages taken over 10 runs, and overall we observed a variance of less than 7%.

#Threads	#States	Time (s)
2	370	0.00
3	13058	0.01
4	356992	0.27
5	8680310	10.76

Table 1. State space sizes and times for Drepper's correct mutex.

Understanding Bugs in Incorrect Variants. Having a formal, checkable model makes it easy to experiment with the intricacies of this futex-based mutex implementation and understand why they are needed. We give two examples of changes to the mutex implementation that compromise its correctness in ways that might not seem immediately obvious. For each, we show that model checking quickly produces short, illuminating counterexample traces.

Bug 1: Incorrect Simplification. On line 8 of the C++ code on page 9, the conditions under which a thread calls `futex_wait` are rather complex and, as discussed by Drepper [7], some of this intricacy is for purposes of optimisation. One might wonder whether, from a correctness point of view, it would suffice for a thread that just failed to lock the mutex to set the futex word to 2 ("locked, waiters"), and call `futex_wait` in an attempt to go to sleep. This would amount to replacing lines 8 and 9 of the C++ code with:

```
futex_word.store(2);
futex_wait(&futex_word, 2);
```

This change does not lead to violations of the `safe_cs` property, but does lead to the possibility of lost waiters. Making corresponding adjustments to `lock()` in our Promela model (including adding a print statement to log the storing of 2 to `futex_word` by a thread), Spin quickly produces the following counterexample when invoked on a 2-threaded configuration:

```
T0 locks mutex on fast path
T1 fails to lock mutex on fast path
T0 decrements futex word from 1 to 0
T0 is done
T1 sets futex.val to 2
T1 futex_wait, value match: 2; sleep
```

The problem is that between T1 observing the mutex to be unavailable and setting the futex word to 2, T0 unlocks the mutex, waking up no waiters, because there are none yet, and terminates. T1 then sets the futex word to 2, goes to sleep and is never woken.

Bug 2: Incorrect `cmpxchg`. On line 10 of the C++ code on page 9, when a thread attempts to lock the mutex on the slow path it tries to change the value of the futex word from 0 to 2, in contrast to the fast path, where a value change from 0 to 1 is attempted (line 6). A reasonable question is: is it essential that the slow path changes the futex word to 2? Adapting the `lock()` implementation in Promela so that the slow path changes the futex word to 1 instead of 2, and applying Spin to a two-threaded configuration leads to successful verification. But with three threads, although `safe_cs` still successfully verifies, Spin quickly reports a counterexample demonstrating an invalid end state:

```
T0 locks mutex on fast path
T1 fails to lock mutex on fast path
T1 futex_wait, value match: 2; sleep
T2 fails to lock mutex on fast path
T2 futex_wait, value match: 2; sleep
T0 decrements futex word from 2 to 1
T0 wakes T2
T0 woke up 1 thread(s)
T0 is done
T2 has woken
T2 locks mutex on slow path
T2 decrements futex word from 1 to 0
T2 is done
```

The counterexample illustrates a situation where threads T1 and T2 go to sleep due to T0 holding the mutex. When the mutex becomes free, T0 wakes up T2, and T0 terminates. T2 then succeeds in locking the mutex on the slow path, but does *not* set the futex word to 2 in the process. As a result, when T2 unlocks the mutex it is not obliged to wake up any waiters, so T1 remains asleep. T2 then terminates, so that T1 becomes a "lost waiter".

This concrete example sheds light on why it is *essential* that the `cmpxchg` used to lock the mutex on the slow path changes the futex word to the "locked, waiters" state: this ensures that if there are additional waiters, the thread that succeeds in locking the mutex on the slow path is guaranteed to wake up one of them. Here model checking facilitates experimenting with design variations, and quickly produces counterexamples that clearly illustrate defects.

6 Model Checking Futex-based Condition Variables

Condition variables (cv) synchronise threads via three operations: cv_wait, cv_signal and cv_broadcast. The cv_wait operation takes a locked mutex as an argument. It atomically unlocks the mutex and puts the calling thread to sleep. Once the thread is woken up, it locks the mutex again before returning. The cv_signal operation wakes up one thread chosen non-deterministically among the sleeping ones, while cv_broadcast, which we ignore hereafter for the sake of conciseness, wakes up all sleeping threads.

The cv_wait operation is atomic in the sense that by the time another thread locks the mutex, the first thread is in the list of threads sleeping on the condition variable. In particular, consider a pair of threads T0 and T1; first T0 holds the mutex and calls cv_wait, then T1 locks the mutex and calls cv_signal: the signal from T1 cannot be *lost*, i.e. it must wake up T0.

6.1 Model Checking Harness and Properties

Like for lock and unlock in Sect. 5, our harness makes use of to-be-defined macros cv_wait and cv_signal, and is designed to have threads loop on calling these two operations while always being able to reach termination. In terms of verification, here we pay special attention to make sure the harness can enable catching *lost signal* bugs by checking freedom from invalid end states.

First, condition variables are used in association with a mutex whose internals are irrelevant, so we define a simple mutex Promela implementation where a mutex is a global boolean variable, the default value of which is false. Locking involves blocking until its value is false before atomically setting it to true, while unlocking simply involves setting it back to false:

```
1 bool mutex;
2 inline mutex_lock() { d_step { !mutex -> mutex = true } }
3 inline mutex_unlock() { mutex = false }
```

The harness consists of a condition variable used by a single *signaller* thread and one or more *waiter* threads. The waiters call cv_wait an arbitrary number of times before terminating. The signaller calls cv_signal until all waiters are done, then it terminates. In order to catch *lost signal* bugs, we also make sure the signaller has an execution path where cv_signal is called only the necessary number of times to match calls to cv_wait, but no more.

To model all this, we start with a constant representing the number of waiters, and a couple of global variables to count the minimum number of signals that are needed and how many threads have terminated, before declaring the waiter threads:

```
1 #define NUM_WAITERS (NUM_THREADS - 1)
2 byte num_signals_req; // Number of signals required
3 byte num_done; // Number of terminated waiter threads
4
5 active[NUM_WAITERS] proctype Waiter() {
6   do
7   :: mutex_lock() ->
8       num_signals_req++;
9       printf("T%d calls cv_wait()\n", _pid);
```

```
10      cv_wait();
11      printf("T%d returns from cv_wait()\n", _pid);
12      mutex_unlock()
13   :: break
14   od
15   num_done++;
16 }
```

Each waiter loops on either locking the mutex, incrementing num_signals_req, calling cv_wait and then unlocking the mutex; or exiting the loop and incrementing num_done before terminating. Thus, each waiter may do an arbitrary number of calls to cv_wait before terminating.

The signaller thread is slightly more complex:

```
1  active proctype Signaller() {
2    do
3    :: num_signals_req > 0 ->
4       mutex_lock();
5       printf("T%d must signal, num_signals_req=%d\n", _pid, num_signals_req);
6       cv_signal();
7       num_signals_req--;
8       mutex_unlock()
9    :: else ->
10      if
11      :: true ->
12         mutex_lock();
13         printf("T%d signals without need\n", _pid);
14         cv_signal();
15         num_signals_req = (num_signals_req > 0 -> num_signals_req - 1 : 0);
16         mutex_unlock()
17      :: true -> printf("T%d won't signal until needed\n", _pid);
18         if
19         :: num_signals_req > 0 -> assert(num_done < NUM_WAITERS)
20         :: num_done == NUM_WAITERS -> assert(num_signals_req == 0); break
21         fi
22      fi
23   od
24 }
```

It loops on either detecting that a signal is required (line 3), in which case it locks the mutex, signals, decrements num_signals_req and unlocks the mutex (lines 4–8); or it sees that no signal is required (line 9). In this case, it non-deterministically decides to either call cv_signal even though there is no apparent need for it (lines 12–16), or to block until either a signal is needed (line 19), or all waiters are done in which case it breaks out of the loop to terminate (line 20). The if branches starting with true (lines 11 and 17) model the "internal" decision of the signaller. In particular, once it has decided to block, it must not signal again unless it detects the need for a signal.

On the one hand, this harness enables the signaller to produce an arbitrary number of signals, even if no waiter is currently waiting for a signal. On the other hand—and this is crucial to detect lost signal bugs—when the signaller sees that no signal is needed, it may decide to stop signalling until either a signal is needed, or all waiters are done. This ensures that each call to cv_wait is matched by at least one call to cv_signal, but potentially no more than the strictly needed number of signals. In the execution path where there is only one signal per wait, if any signal is lost this will lead to a scenario where (a) some waiter is stuck in the cv_wait call at line 10, and (b) the signaller is blocked at line 19 because no signals are currently required. Thus the lost signal will lead to the model checker reporting an invalid end state.

The rest of this section covers a couple of futex-based implementations of cv_wait and cv_signal, as presented in [6]. Each implementation requires a single futex, which is always declared as a global variable named futex.

6.2 Take 1: Naive and Incorrect

We start with a naive approach, from the "Simple but very wrong" section in [6]:

```
1  class CondVar {
2  public:
3    CondVar() : futex_word(0) {}
4    void cv_wait(mutex &m) {
5      m.unlock();
6      futex_wait(&futex_word, 0);
7      m.lock();
8    }
9    void cv_signal() { futex_wake(&futex_word, 1); }
10
11 private:
12   atomic<uint32_t> futex_word;
13 };
```

The cv_wait operation unlocks the mutex before calling futex_wait with a plain value of 0 (the initial value of the futex word) to put the thread to sleep. Upon waking up, it locks the mutex again before returning. The cv_signal operation just calls futex_wake to wake up one of the sleeping threads.

This is modelled in Promela using the following macros:

```
1 inline cv_wait() {
2   mutex_unlock();
3   futex_wait(futex, 0);
4   mutex_lock();
5 }
6
7 inline cv_signal() { futex_wake(futex, 1) }
```

Invoking Spin on the harness with this version leads to an invalid end state error. Spin produces a counterexample that illustrates the issue: after the mutex is unlocked in cv_wait (line 2), the signaller thread might call cv_signal and thus futex_wake before the waiter calls futex_wait (line 3); the signal is lost. In this case, if the signaller decides to block until another signal is needed, then the waiter thread has no chance to be woken up: the system is in deadlock.

6.3 Take 2: Bionic, Unlikely yet Possible Deadlock

Our second take, dubbed "Sequence counter, close but no cigar" in [6], mimics Android's Bionic libc [1] approach to implement condition variables, where cv_signal increments the futex word to avoid deadlocks seen in take 1:

```
1 class CondVar {
2 public:
3   CondVar() : futex_word(0) {}
4   void cv_wait(mutex &m) {
5     uint32_t old_value = futex_word;
6     m.unlock();
7     futex_wait(&futex_word, old_value);
```

```
8      m.lock();
9    }
10   void cv_signal() {
11     futex_word.fetch_add(1);
12     futex_wake(&futex_word, 1);
13   }
14
15 private:
16   atomic<uint32_t> futex_word;
17 };
```

In `cv_wait`, the value of the futex word is saved in `old_value` before releasing the mutex, then `futex_wait` is called with `old_value`. In `cv_signal`, the futex word is incremented by 1, with a possible overflow, before calling `futex_wake`. This avoids the deadlock situation encountered in Sect. 6.2: if `cv_signal` is executed between unlocking the mutex (line 6) and calling `futex_wait` (line 7) in `cv_wait`, the futex word value will be different from the value used in the call to `futex_wait` which thus will not block.

This is modelled in Promela using the following macros:

```
1  inline cv_wait() {
2    byte val = futex.word;
3    mutex_unlock();
4    futex_wait(futex, val);
5    mutex_lock();
6  }
7
8  inline cv_signal() {
9    futex.word = inc(futex.word);
10   futex_wake(futex, 1);
11 }
```

However, Spin still reports a possible deadlock: if between lines 3 and 4, `cv_signal` is called enough times to overflow the futex word and bring it back to the `old_value` saved in line 2, then the call to `futex_wait` does block, and we reach a deadlock. This issue is documented in Bionic, with an acknowledgement that it would be extremely unlikely to arise in practice: with a 32-bit futex word, we would need *exactly* 2^{32} calls to `cv_signal` in a row, while `cv_wait` is between lines 3 and 4, to trigger the deadlock.

Such issues are hard to foresee at design time. Model checking is valuable in illustrating rare risks of deadlocks, and evaluating their acceptability in practice.

7 Related Work

There is a significant literature on formal verification of inter-process communication primitives. Bogunovic *et al.* verified mutual exclusion algorithms with SMV [4], with an analysis of liveness and fairness. Mateescu and Serwe analysed 27 different shared-memory mutual exclusion protocols with CADP for both correctness and performance [15,16]. Bar-David and Taubenfeld used model checking techniques to automatically discover mutual exclusion algorithms [2]. More recently, Kokologiannakis and Vafeiadis developed a specific dynamic partial order reduction (DPOR) technique to better handle the barrier synchronisation primitive [14]. In terms of using model checking for education, Hamberg and Vaandrager wrote about their experience using UPPAAL in a course on operating systems [12].

We are not aware of formal verification of futex-based synchronisation primitives. Futexes are primarily a Linux system call [10, 11]. Besides the two reference publications from Franke *et al.* [9] and Drepper [7], Benderski wrote a good introduction on the topic [3]. Note that the futex system call itself has suffered from bugs that affected userspace applications, such as the Java VM [17].

8 Future Directions

We have presented a case study of modelling a series of futex-based implementations of mutexes and condition variables in Promela, and using Spin to verify safety properties. An immediate extension would be to consider fairness to enable verifying liveness properties, like the absence of starvation. We can also explore additional futex-based synchronisation primitives, for instance barriers.

To create an educational resource that would require little model checking expertise, we can think of doing verification directly on C implementation by using a C model checker, like CBMC [5]. We can even envision extracting C models from various C standard library implementations (e.g. glibc), to verify designs actually used in widespread libraries. Finally, it would be interesting to verify the implementation of the futex system call implementation itself in the Linux kernel and other OSes that have adopted futexes (e.g. OpenBSD).

References

1. Android: Bionic C library, pthread_cond implementation (2023). https://android.googlesource.com/platform/bionic/+/refs/tags/android-13.0.0_r24/libc/bionic/pthread_cond.cpp. Accessed 10 Jan 2023
2. Bar-David, Y., Taubenfeld, G.: Automatic discovery of mutual exclusion algorithms. In: Fich, F.E. (ed.) DISC 2003. LNCS, vol. 2848, pp. 136–150. Springer, Heidelberg (2003). https://doi.org/10.1007/978-3-540-39989-6_10
3. Benderski, E.: Basics of futexes (2018). https://eli.thegreenplace.net/2018/basics-of-futexes/. Accessed 10 Jan 2023
4. Bogunovic, N., Pek, E.: Verification of mutual exclusion algorithms with SMV system. In: The IEEE Region 8 EUROCON 2003. Computer as a Tool. vol. 2, pp. 21–25. IEEE (2003)
5. Clarke, E., Kroening, D., Lerda, F.: A tool for checking ANSI-C programs. In: Jensen, K., Podelski, A. (eds.) TACAS 2004. LNCS, vol. 2988, pp. 168–176. Springer, Heidelberg (2004). https://doi.org/10.1007/978-3-540-24730-2_15
6. Denis-Courmont, R.: Condition variable with futex (2020). https://www.remlab.net/op/futex-condvar.shtml. Accessed 10 Jan 2023
7. Drepper, U.: Futexes are tricky (2011). https://www.akkadia.org/drepper/futex.pdf. Accessed 10 Jan 2023
8. Evrard, H., Donaldson, A.: Model checking futexes: code examples (2022). https://github.com/mc-imperial/modelcheckingfutexes. Accessed 16 Jan 2022
9. Franke, H., Russell, R., Kirkwood, M.: Fuss, futexes and furwocks: fast userlevel locking in Linux. In: Ottawa Linux Symposium 2002, pp. 479–495 (2002). https://www.kernel.org/doc/ols/2002/ols2002-pages-479-495.pdf. Accessed 10 Jan 2022

10. Futex manual page section 2 (system calls) (2023). https://man7.org/linux/man-pages/man2/futex.2.html. Accessed 16 Jan 2023
11. Futex manual page section 7 (miscellaneous) (2023). https://man7.org/linux/man-pages/man7/futex.7.html. Accessed 16 Jan 2023
12. Hamberg, R., Vaandrager, F.: Using model checkers in an introductory course on operating systems. ACM SIGOPS Operat. Syst. Rev. **42**(6), 101–111 (2008)
13. Holzmann, G.: The SPIN model checker: primer and reference manual. Addison-Wesley Professional, 1st edn. (2011)
14. Kokologiannakis, M., Vafeiadis, V.: BAM: efficient model checking for barriers. In: Echihabi, K., Meyer, R. (eds.) NETYS 2021. LNCS, vol. 12754, pp. 223–239. Springer, Cham (2021). https://doi.org/10.1007/978-3-030-91014-3_16
15. Mateescu, R., Serwe, W.: A study of shared-memory mutual exclusion protocols using CADP. In: Kowalewski, S., Roveri, M. (eds.) FMICS 2010. LNCS, vol. 6371, pp. 180–197. Springer, Heidelberg (2010). https://doi.org/10.1007/978-3-642-15898-8_12
16. Mateescu, R., Serwe, W.: Model checking and performance evaluation with CADP illustrated on shared-memory mutual exclusion protocols. Sci. Comput. Program. **78**(7), 843–861 (2013)
17. Mechanical sympathy email group, discussion titled linux futex_wait() bug (2015). https://groups.google.com/g/mechanical-sympathy/c/QbmpZxp6C64. Accessed 16 Jan 2023

Sound Concurrent Traces for Online Monitoring

Chukri Soueidi(✉) and Yliès Falcone

Univ. Grenoble Alpes, Inria, CNRS, Grenoble INP, LIG, 38000 Grenoble, France
{chukri.a.soueidi,ylies.falcone}@inria.fr

Abstract. Monitoring concurrent programs typically rely on collecting traces to abstract program executions. However, existing approaches targeting general behavioral properties are either not tailored for online monitoring, are no longer maintained, or implement naive instrumentation that often leads to unsound verdicts. We first define the notion of when a trace is representative of a concurrent execution. We then present a non-blocking vector clock algorithm to collect sound concurrent traces on the fly reflecting the partial order between events. Moreover, concurrent events in the representative trace pose a soundness problem for monitors synthesized from total order formalisms. For this, we extract a causal dependence relation from the monitor to check if the trace has the needed orderings and define the conditions to decide at runtime when a collected trace is monitorable. We implement our contributions in a tool, FACTS, which instruments programs compiling to Java bytecode, constructs sound representative traces, and warns the monitor about non-monitorable traces. We evaluate our work and compare it with existing approaches.

1 Introduction

Guaranteeing the correctness of concurrent programs often relies on dynamic verification which requires observing and abstracting the program behavior. Abstraction is typically provided by *traces* that contain the executed actions which can be analyzed online or post-mortem. Traces serve as a model for property-based *detection* and *prediction* techniques which choose their trace collection approaches differently based on the class of targeted properties. Some approaches target generic classic concurrency errors such as data-races [31,37], deadlocks [14], and atomicity violations [30,53,72]. Other techniques target general behavioral properties; those are typically order violations such as null-pointer dereferences [29], and typestate violations [36,38,65] and more generally runtime verification [27,47].

When properties require reasoning about concurrency in the program, causality between events must be established during trace collection. Data race detection techniques for instance require it to check for concurrent accesses to shared variables; as well as predictive approaches targeting behavioral properties such as [22,37,38,65] in order to explore other feasible executions. Causality is best

G. Caltais and C. Schilling (Eds.): SPIN 2023, LNCS 13872, pp. 59–80, 2023.
https://doi.org/10.1007/978-3-031-32157-3_4

expressed as a partial order over events, which are compatible with various formalisms for the behavior of concurrent programs such as weak memory consistency models [4,7,52], Mazurkiewicz traces [33,55], parallel series [48], Message Sequence Charts graphs [56], and Petri Nets [57].

However, while a concurrent program behaves non-sequentially, trace collection is sequential and the causality between events must be established during or after observation. For many monitoring approaches, the program is a black box and the trace is the sole system information provided. Thus, providing traces with correct and sufficient ordering information is necessary for sound and expressive monitoring. Any *representative* trace for a concurrent execution must only include correct information and must reflect all the ordering information needed to verify a behavioral property. Thus, it is best modeled as a partial order over program actions.

This paper addresses the problem of providing representative traces for existing verification techniques to soundly monitor concurrent programs *online*. We focus on general *behavioral* properties targeting violations that cannot be traced back to classical concurrency errors. These properties are usually expressed using total-order formalisms such as LTL and finite-state machines [1,50]. One example is a precedence property, which specifies that a resource can only be granted (event g) in response to a request (event r), and can be expressed in LTL as $\neg g \mathbf{W} r$. Unfortunately, these properties have received less attention in the literature in the context of concurrent programs.

Existing approaches that monitor such properties are either not tailored for online monitoring, no longer maintained, or implement naive instrumentation. Let us first consider dynamic *predictive* techniques that target behavioral properties, such as [22,37,38,65]. These techniques often rely on vector clock algorithms to timestamp events, construct partial orders and build causal models to explore feasible executions. Although they capture representative traces, they are expensive to use online and often intended for the design phase of the program and not in production environments (see Sect. 8). Moreover, by serializing reads and writes in the program, they assume a sequentially consistent execution model [35]. For simple detection, this forces a specific schedule on the execution, and for prediction, it restricts the space of possible interleavings. Furthermore, such techniques are incomplete and cannot prove the absence of property violations (they never claimed it) which is a reason why online monitoring is inevitable for many systems.

We turn to classical monitoring (*detection*) techniques and we find that many top existing approaches (based on the competitions on RV [11,28,61]) were initially developed to monitor single-threaded programs and later adapted to multithreaded programs. Unfortunately, as shown in [25], these approaches problematically impose a total order on events using incorrect assumptions. They have wrongly taken for granted that the instrumentation always yields representative traces. When events of interest execute in concurrent regions, these tools often produce unsound and inconsistent verdicts. These approaches can leverage representative traces to produce sound verdicts in the presence of concurrent events, and their expressiveness can be extended to check concurrency-related behavioral properties. For example, one might be interested to verify the correctness of the scheduling algorithms of tasks with data dependencies (e.g., [71]).

We first revisit traces and qualify two properties that determine if they are good representatives of an execution: *soundness* and *faithfulness* (Sect. 4). Soundness holds when the trace does not provide false information about the order of events. Faithfulness holds when the trace contains all the information about the order of events from the execution. **RQ1:** *Can we collect a representative trace of a concurrent execution on the fly with minimal interference on the program?* Vector clock algorithms have been employed and refined for several decades now [5,18,53,63]. However, very few (we know of [62]) are directed towards online monitoring of behavioral properties; where a final trace consists of property-related events only. These algorithms typically require blocking the execution; by synchronizing the instrumentation, program actions, and algorithm's processing to avoid data races [18]. With the quadratic bound on their runtime complexity and the coarse-grained synchronization they introduce, one might want to run such an algorithm off the critical path of the program. We present a vector clock algorithm (Sect. 5) that does not require blocking the execution and can run either synchronously or asynchronously with the program. Asynchronous trace collection is ideal for scenarios where the monitoring overhead cannot be afforded and a small delay in the verdict can be tolerated. For instance, in real-time systems where the system is expected to produce a result within a defined strict deadline [66]. Our algorithm constructs representative concurrent traces that are sound and very often faithful[1]. As far as we know, it is unique in the context of monitoring behavioral properties that can run off the critical path of the execution.

RQ2: *Is the collected concurrent trace good enough to soundly monitor a property?* Monitoring single-threaded programs depend on instrumentation, which is assumed to be correct, to provide all relevant events [12]. For concurrent programs, we notice that monitorability with classical approaches depends also on the ordering information available in the trace (resp. the execution). Firstly, consider for instance the precedence property seen previously. If events r and g happen to execute concurrently in the program, a sound trace would have them as unordered events and a sound monitor should report a violation of the property. However, in practice, when monitoring such property with an automaton, the partial order must be linearized before passing it to the monitor. The linearization will produce an arbitrary order between r and g, for instance, $r.g$ which would make the oblivious monitor miss the violation. In such a case, the trace is not fit for monitoring the property and the monitor should be warned. Secondly, in certain scenarios due to some partial instrumentation or logging failure, some synchronization actions might be missing from the trace, resulting in unfaithful traces where some orderings cannot be established. This also poses soundness problems to the monitor similar to the problems with lossy

[1] The algorithm might miss, in some marginal cases, ordering information resulting in sound but unfaithful traces (see Sect. 5).

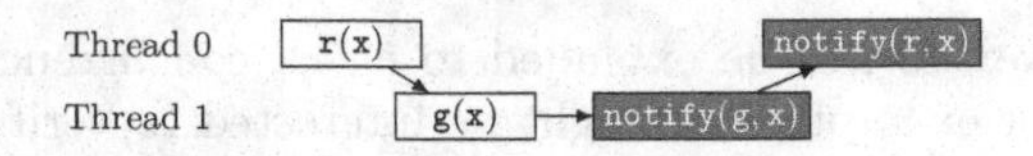

Fig. 1. Unsound instrumentation.

traces [39]. To handle the mentioned problems, we extract a *causal dependence relation* from a given property to know which events cannot permute in a trace and check whether a trace contains enough order information (Sect. 6). We then redefine trace *monitorability* for concurrent executions with a necessary condition on the trace to guarantee a sound verdict when monitoring. If the condition is not met, we produce warnings for the monitor.

In this paper we target shared memory concurrency, however, our work can be adapted to message passing. We implement our contributions in a tool, FACTS, which attaches to programs running on the JVM, instruments and collects sound and faithful traces, and checks the monitorability criteria. We evaluate our approach and demonstrate its effectiveness in capturing concurrent traces and provide a cost estimation on the execution time overheads (Sect. 7). We also implement the algorithm from [63], used in [22,38,64], and show how our algorithm has better advantages.

2 Issues with Linear Traces

Monitoring approaches relying on total-order formalisms such as LTL and finite state machines require linear traces to be fed to the monitors as their input consists of words. One might be tempted to immediately capture linear traces from a concurrent execution without reestablishing the causal order. In this section, we discuss issues of collecting linear traces to motivate concurrent traces.

Advice Atomicity Assumption. General purpose runtime verification frameworks such as Java-MOP [21], Tracematches [8], and others [11] rely on instrumentation for extracting events for traces. To handle concurrent programs, RV tools provide a feature to *synchronize* the monitor to protect it from concurrent access and data races. As such, threads acquire a lock before notifying the monitor and release it afterward. However, when a program action is executed, its advice will not always execute atomically with it unless both are wrapped in a synchronization block. Consider Fig. 1 and the property `r` *precedes* `g`. A context switch happens at $t1$ after the `r(x)` is executed (but before notifying the monitor), allowing the execution of `g(x)` and its advice before `notify(r,x)`. It is clear how this would result in unsound monitoring. This observation problem cannot be solved by simply checking for the absence of data races. One needs to guarantee atomicity between all executing program actions and their advice. However, the problem will not appear if `r` and `g` are synchronized and executed with their advice within mutually exclusive regions. Nevertheless, we might sometimes want to capture events from concurrent regions.

Forced Atomicity. One way to solve the lack of advice atomicity is to force it. Forced atomicity can be achieved by instrumenting synchronization blocks[2] that wrap the program actions with their advice in mutually exclusive regions. However, forcing atomicity introduces two new problems, one at the program level and the other at the monitor level. First, it forces a total order between concurrent program actions; interfering with the parallelism of the program and changing its behavior. One needs also to minimize the area for which the lock is applied and avoid coarse-grained synchronization. From the monitor side, the verdict will be dependent on the specific scheduling of the execution. Take `r` *precedes* `g` for instance, if these events are concurrent in the program, monitoring will produce a different verdict at each run. Although the verdict is correct, we would like to warn the user. Second, any information about parallel actions in the program is lost and one can no longer determine whether two actions execute concurrently initially in the non-instrumented program. In that case, it becomes impossible to express properties on the concurrent parts of the execution. Furthermore, RV tools mentioned above rely on AspectJ [40] for a high-level specification of instrumentation which is rather unfitting to instrument synchronization blocks as it cannot instrument at the bytecode level.

3 Characterizing a Concurrent Execution

We first choose the smallest observable execution step done by a program, such as a method call or write operation, and refer to this step as an *action*. An *action* is a tuple of attribute-value pairs. Common attributes are *id*, associated *resource* i.e. shared variable, integer *id* to distinguish different actions corresponding to the same instruction[3]. Actions executing in the same thread follow a *thread order*. The set of all program actions is denoted by $\mathbb{A}$. We distinguish between two types of actions: *regular actions* and *synchronization actions*. Regular actions are relevant to the property we want to monitor. Synchronization actions, denoted by the set SA, are used for synchronization between different threads. They provide *release-acquire* ordering semantics [3] and establish a *synchronization order* on the execution. Essentially, an acquire action a' on a resource by some thread synchronizes with the latest release action a, if it exists, on that same resource by some other thread. Given two threads t and u, we highlight the following synchronization actions that can establish release-acquire:

- $\mathit{unlock}(t,l)/\mathit{lock}(t,l)$: release/acquire of lock l by t;
- $\mathit{fork}(t,u)/\mathit{begin}(u)$: fork of u by t/first action by u;
- $\mathit{end}(t)/\mathit{join}(u,t)$: last action t/ u blocking until t ends;
- $\mathit{write}(t,x,v)/\mathit{read}(t,x,v)$: value v on shared variable x.
- $\mathit{notify}(t,s)/\mathit{wait}(t,s)$: notify/wait a signal s.

[2] Such as `synchronized` in Java.
[3] We depict actions using the notation $\texttt{id.label}^{\texttt{tid}}_{\texttt{resource}}$ in diagrams.

The *execution order* $\xrightarrow{e} \subseteq (\mathbb{A} \times \mathbb{A})$ is the transitive closure of both *thread* and *synchronization* orders. This order between actions is a partial order[4] and is general enough to represent various formalisms and models of concurrent systems.

Definition 1 (Concurrent Execution). *A concurrent execution is a partially-ordered set of actions* $(\mathbb{A}, \xrightarrow{e})$.

4 Sound and Faithful Concurrent Traces

Our goal is to capture a representative trace from a concurrent execution. In [67], we introduce two fundamental notions for concurrent traces: *soundness* and *faithfulness*. We first revisit both notions here, then proceed to show how such traces preserve the properties of an execution. We note that the trace actions $\mathcal{E}$ are often called *event* in monitoring and verification approaches.

Definition 2 (Concurrent Trace). *A concurrent trace is a partially ordered set* $t = (\mathcal{E}, \xrightarrow{tr})$ *of events such that* $\mathcal{E} \subseteq \mathbb{A}$.

Informally, a trace is a sound trace if it does not provide false information about the execution.

Definition 3 (Trace Soundness). *A concurrent trace t is said to be a sound trace of a concurrent execution e (noted* $\mathrm{snd}(e, t)$*) iff (i)* $\mathcal{E} \subseteq \mathbb{A}$ *and (ii)* $\xrightarrow{tr} \subseteq \xrightarrow{e}$.

To be sound, a trace (i) should not capture an action not found in the execution, and (ii) should not order unrelated actions.

While a trace that does not provide incorrect information about the execution model leads to sound monitoring, a trace can still not provide enough information about the order (for a monitor to determine a verdict). Faithfulness is similar to *completeness*; it is expensive as it requires capturing all relevant causality in the program. Informally, a *faithful* trace contains all information on the order of events that occurred in the execution model.

Definition 4 (Trace Faithfulness). *A concurrent trace t is said to be faithful to an execution e (noted* $\mathrm{faith}(e, t)$*) iff* $\xrightarrow{tr} \supseteq (\xrightarrow{e} \cap\, \mathcal{E} \times \mathcal{E})$.

Example 1 (Soundness and Faithfulness). Consider an execution of a properly synchronized Java program with three threads, a writer, and two readers, where the reader threads performed a concurrent read in between two write operations. We omit the depiction of actions performed by each thread such as locks, etc., for brevity. Four different collected traces from such execution are presented in Fig. 2. For a behavioral property such as "no read or write happen concurrently with another read or write", we are interested in the actions `r` and `w`. The order relative to these events in the execution is $\rightarrow_{e_0} = \{\langle 0.\mathtt{w}^0, \mathtt{r}^1\rangle, \langle 0.\mathtt{w}^0, \mathtt{r}^2\rangle, \langle \mathtt{r}^1, 1.\mathtt{w}^0\rangle, \langle \mathtt{r}^2, 1.\mathtt{w}^0\rangle, \langle 0.\mathtt{w}^0, 1.\mathtt{w}^0\rangle\}$. Figure 2a presents

[4] We assume familiarity with partial orders and partially ordered sets.

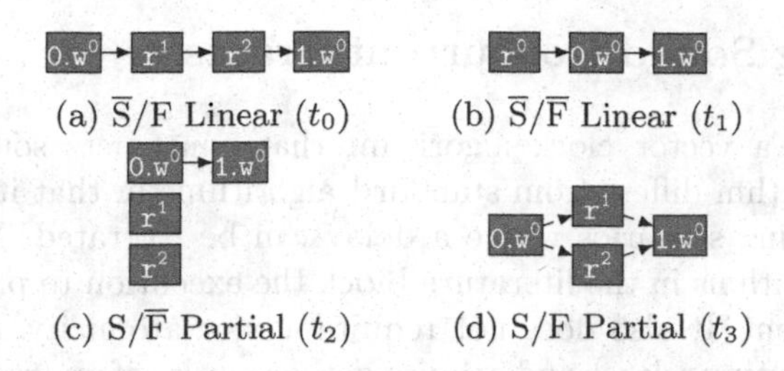

(a) $\overline{\text{S}}$/F Linear (t_0) (b) $\overline{\text{S}}/\overline{\text{F}}$ Linear (t_1)

(c) S/$\overline{\text{F}}$ Partial (t_2) (d) S/F Partial (t_3)

Fig. 2. Four different collected traces from the execution of 1-Writer 2-Readers.

a linear trace of the execution t_0 as captured by Java-MOP using advice atomicity assumption, see Sect. 2. One can see that the order is $\rightarrow_{t_0} = \rightarrow_{e_0} \cup \{\langle \mathtt{r}^1, \mathtt{r}^2 \rangle\}$. We notice that we have faithfulness ($\mathrm{faith}(e_0, t_0)$) as $\rightarrow_{e_0} \subset \rightarrow_{t_0}$. However we do not have soundness ($\neg\mathrm{snd}(e_0, t_0)$) as the pair $\langle \mathtt{r}^1, \mathtt{r}^2 \rangle \notin \rightarrow_{e_0}$, as the reads happened concurrently. Figure 2b shows a trace that is neither sound nor faithful. Figure 2c the trace captures thread order only. It is a sound trace, as it only contains $\langle \mathtt{0.w}^0, \mathtt{1.w}^0 \rangle$, and therefore no wrong information. However, it is not faithful, as it is missing order information. Figure 2d presents a partial trace of the execution t_3 that is both sound and faithful. Ideally, t_3 is the smallest concurrent trace collected to verify behavioral properties on reads and writes.

A property can be used to define the set of correct concurrent executions. For the verification of the temporal behavior of programs, the semantics of matching properties is applied on traces [51]. Effectively, at the semantic level, a property partitions the set of all executions into correct and incorrect ones. Consequently, a property $\varphi(\Sigma)$ defined over $\Sigma \subseteq \mathbb{A}$ is a set of partial orders over Σ.

Definition 5 (Property Satisfaction). *A concurrent execution* $(\mathbb{A}, \xrightarrow{\mathrm{e}})$ *satisfies a property* $\varphi(\Sigma)$ *(noted* $(\mathbb{A}, \xrightarrow{\mathrm{e}}) \models \varphi(\Sigma)$*) iff* $(\xrightarrow{\mathrm{e}} \cap \Sigma \times \Sigma) \in \varphi(\Sigma)$.

To check that a property has been satisfied, we simply "project" the order $\xrightarrow{\mathrm{e}}$ on Σ, that is, we restrict our information about the execution to Σ. We then check if the projection belongs to the set of correct concurrent executions ($\varphi(\Sigma)$).

While our goal is verifying properties on the full execution of a program, we generally gather a subset of it as a trace. As such, we are interested in the fact that verifying a property on the trace holds the same as it would on the full execution. By construction (from Definition 3 and 4), we notice that the projections over Σ (Definition 5) of some execution e and a trace t for a property over variables $\Sigma \subseteq \mathcal{E} \subseteq \mathbb{A}$ where we have soundness ($\mathrm{snd}(e, t)$) and faithfulness ($\mathrm{faith}(e, t)$) are the same. We deduce the following theorem.

Theorem 1 (Property Preservation). *Given a concurrent execution* $\mathrm{e} = (\mathbb{A}, \xrightarrow{\mathrm{e}})$, *a trace* $\mathrm{t} = (\mathcal{E}, \xrightarrow{\mathrm{tr}})$, *and a property* φ *over* $\Sigma \subseteq \mathcal{E} \subseteq \mathbb{A}$ *we have:*

$$(\mathrm{snd}(\mathrm{e}, \mathrm{t}) \wedge \mathrm{faith}(\mathrm{e}, \mathrm{t})) \implies (\mathrm{e} \models \varphi(\Sigma) \text{ iff } \mathrm{t} \models \varphi(\Sigma)).$$

We say that t is an appropriate abstraction of e; e and t can be used interchangeably to verify properties over Σ. Since our notion of a property is simply a set of traces, the presented results apply to any set of monitorable [26,58] properties.

5 Obtaining Sound Concurrent Traces

We here present a vector clock algorithm that constructs sound concurrent traces. The algorithm differs from standard algorithms in that it can run asynchronously, allowing scenarios where a delay can be tolerated. Most, if not all vector clock algorithms in the literature block the execution to process the algorithm at each event. It also does not require advice atomicity. Instead, it just requires instrumenting release and acquire actions with *before* (resp. *after*) directives. When an action a is instrumented with *before* (resp. *after*) direction, the notification action, `notify(a)`, should execute before a (resp. *after*), and no other action from the same thread should execute in between.

Let us first review notions of vector clocks [44, 54]. A vector clock is a function from threads to integers, $V : T \rightarrow Int$ with three key operations. *Join* is the element-wise maximum of two vector clocks: $V_1 \sqcup V_2 = V_3$ such that $\forall t : V_3(t) = max(V_1(t), V_2(t))$. *Comparison* is the element-wise comparison of two vector clocks: $V_a \leq V_b$ is defined to mean $\forall t.\ V_a(t) \leq V_b(t)$. Finally, function *inc* increases a single component of a vector clock: $inc_t(V) = \lambda u.(\text{if } u = t \text{ then } V(u) + 1 \text{ else } V(u))$.

5.1 The Reordering Algorithm

Algorithm 1 maintains the following data structures. A map $\mathbb{L} : T \rightarrow V$ holding the last timestamp seen by each thread. A map $\mathbb{R}$ holding the last release action per resource, a map $\mathbb{W}$ holding the last write of a value on a shared variable, and a collection $\mathbb{T}$ which represent the concurrent trace and will be populated with timestamped actions. On each received notification action, the algorithm sets its timestamp to the latest timestamp seen by its thread (line 3). Then, synchronization actions in SA (lines 4-5) are send to `ReleaseOrAcquire`, except for *read* or *write*. Actions *unlock*, *fork*, *end* are represented with *release*, and the algorithm puts them in the map entry in $\mathbb{R}$ associated with their resource in R (lines 11-12). Actions *lock*, *begin*, *join* are represented with *acquire*, the algorithm retrieves the last action that released its resource from R; if found[5], their vector clocks are merged (lines 13-15). The *join* is merged with the last action seen by the finished thread. Actions *read* and *write* are handled with `ReadOrWrite`. They are handled with the map $\mathbb{W}$ which is indexed by a shared variable and a value. A write (lines 17-23) is only pushed into $\mathbb{W}$ when it is not conflicting with the entry in $\mathbb{W}$ associated with its variable and value (discussed more in the following paragraph). A read is merged with the latest write (lines 24-26). After that, the timestamp is incremented (line 6), and map $\mathbb{L}$ is updated (line 7). Events will be stored in the concurrent trace, and all synchronization actions will be discarded (lines 8-9).

The *Join*, *Comparison* and *Copy* operations of vector clocks require $\Theta(k)$ time, linear in the number of threads k. *Increment* operations on vector clocks,

[5] We omit some null checks to simplify the presentation of the algorithm, however, assume that joining $e.VC$ with a null value does not affect it.

Algorithm 1: Vector Clock Algorithm

```
1  procedure ReceiveAction(e)
2      t = tid(e)
3      e.VC = copy(𝕃(t))
4      if e is a synchronization action then
5          send e to the appropriate procedure
6      inc_t(e.VC)
7      𝕃(t) = e.VC
8      if e is not a synchronization action then
9          𝕋 ← e                                  // Add e to trace 𝕋
10 procedure ReleaseOrAcquire(e)
11     if e = release(t, r) then
12         ℝ(r) = e                               // Update ℝ
13     else if e = acquire(t, r) then
14         e′ = ℝ(r)
15         e.VC = e.VC ⊔ e′.VC                    // Merge vector clocks
16 procedure ReadOrWrite(e)
17     if e = write(t, x, v) then
18         e′ = 𝕎(x, v)
19         if (e′ = null) ∨ (e′.VC ≤ e.VC) then
20             𝕎(x, v) = e
21         else
22             𝕎(x, v) = null                     // Conflicting write
23         end
24     else if e = read(t, x, v) then
25         e′ = 𝕎(x, v)
26         e.VC = e.VC ⊔ e′.VC
```

retrieving and inserting elements to the maps require $O(1)$. The reordering algorithm then requires $O(n \times k)$ time for a trace of n actions.

5.2 Correctness Discussion

We now show how the algorithm always produces sound traces. After the algorithm ends, the collected linear trace is timestamped into a concurrent trace represented by $\mathbb{T}$ that contains ordered pairs $\langle a, b\rangle$, such that $a.VC \leq b.VC$. For handling *release* and *acquire* actions (except for reads and writes), these actions already execute in a total order in any execution. The algorithm performs classical operations of timestamp merging for matching actions. However, the correctness of our algorithm is dependent on instrumentation. Since we do not force atomicity between an action and its notification, we must instrument actions that perform vector clock merge operations i.e. *acquire* actions with the *after* directive and *release* actions with the *before* directive. The intuition is that if a *release* action is instrumented with the *after* directive, a context switch between it and its advice can lead to having an unmerged consecutive *acquire*.

With proper instrumentation, whenever we observe an *acquire* action, we are sure that if a matching *release* exists, then it has been already been captured and processed. When the program only uses *lock* and *fork* actions for synchronization, then our algorithm guarantees both soundness and faithfulness of the concurrent trace since it captures all orderings, provided that they are all instrumented and captured.

Handling *reads* and *writes* is more demanding if we want to run the algorithm asynchronously and do not want to force on them a total order in the execution. Reads and writes are instrumented with *after* and *before* directives, respectively, including volatile and atomic variables [46][6]. When observing a read r on shared variable x with a value v (line 24), we know its matching write was processed. Nevertheless, if prior to r, two (or more) threads performed writes, w and w', on x with the same value v, then we need to make sure that w and w' are ordered since r could be reading the value written by any one of them; hence its clock cannot be safely merged with any of them. We say the two writes on x, writing the same value v, are *conflicting* when they are not ordered, i.e. $w.VC \not\leq w'.VC$ and $w'.VC \not\leq w.VC$. On a write event, the algorithm always compares it with the previously observed write on the shared variable. If the previous write event had written the same value and the new and previous writes are not ordered (line 19), then we clear the map entry holding the last write (line 22) so that a future read cannot merge with it. To guarantee the correctness of the matching, we push the last write on x into $\mathbb{W}$ if there are no conflicting writes (line 20). Hence, a *read* will either match its write and merge, or not. This might have consequences on the faithfulness of the trace and not its soundness. Faithfulness will only be affected if (1) three or more threads are writing to the same variable the same value and in a concurrent region, and (2) the program relies on those writes to synchronize. We reasonably believe that this is an infrequent case in concurrent programs. However, for such marginal cases, the algorithm can report conflicting writes. Writes can then be instrumented to have advice atomicity forcing a total order on their execution; by adding synchronization blocks that wrap them with their advice in mutually exclusive regions. When this applies, there will be no need for the procedure `ReadOrWrite`, and reads and writes can be treated as *acquire* and *release* where they are handled with procedure `ReleaseOrAcquire`.

6 Criteria for Monitorability

In this section, we discuss the criteria for sound monitoring with automata and concurrent traces. In Sect. 4, we see that a sound and faithful trace represents a program execution and can be used interchangeably when verifying a property. However, concurrent traces (equivalently concurrent executions) might have unordered events. Many monitoring approaches rely on finite state automata as they can be used for most of the specification pattern [1,24]. Such monitors

[6] Atomic operations such as *compare-and-swap* are handled differently since we check for the result of the operation before emitting an action.

expect a total order of events as their input consists of words. Therefore, concurrent traces must be linearized before monitoring. A *linearization* of a partial order will impose an arbitrary order between these unordered events. However, feeding the monitor with incorrect orderings of events can lead to incorrect verdicts. For the remainder of this section, we set Σ as the set of events over which properties are specified. We distinguish it from $\mathcal{E}$, which is the set of runtime events that will possibly be projected to events in Σ.

6.1 Monitor Causal Dependence

When observing an automaton, we might find pairs of events whose order is irrelevant to its progress; they can permute without affecting the verdict. The *causal dependence* relation $\mathcal{D} \subseteq \Sigma \times \Sigma$ is a binary relation that is anti-reflexive and symmetric. It contains all pairs of events whose correct order is necessary and their permutation would lead the automaton to a different state. Hence, for all pairs of events that do not belong to $\mathcal{D}$, their permutation after a linearization (if they occur concurrently) can be safely tolerated. Moreover, $\mathcal{I}_\mathcal{D} = (\Sigma \times \Sigma) \setminus \mathcal{D}$ is the independence relation induced by the dependence relation $\mathcal{D}$. For a deterministic automaton, $\mathcal{I}_\mathcal{D}$ is a reflexive and symmetric relation. We extract $\mathcal{I}_\mathcal{D}$ by finding pairs of actions that would lead to the same state if permuted. An algorithm that extracts $\mathcal{I}_\mathcal{D}$ from a property specified as a DFA is fairly simple (see technical report [68]). It checks all pairs (a, b) in $\Sigma \times \Sigma$, if starting from any state in the automaton $a.b$ would lead to a different state than $b.a$. If so, then the automaton depends on receiving both symbols in order, and the pair is added to the dependency relation. Note that, the *causal dependence* relation resembles trace equivalence from [55] which constrains the allowed linearizations of a partial order. However, here we extract the relation from the monitor itself, and it defines *word* equivalence concerning an automaton, that is, the allowed permutations of letters in a word that would eventually lead to the same verdict.

Example 2 (Monitor Causal Dependence). We demonstrate with two example properties[7]. **P1** states that event s responds to p between q and r. **P2** is a mutual exclusion property that states that no read or write should happen concurrently with another write. To monitor with a finite state machine monitor, the read and writes are instrumented and delimited with events *bw* and *aw* represent before and after a write, *br* and *ar* for before and after a read. Figure 3 shows the violation automata of bad prefixes to monitor the properties. For each automaton, we have:

- $\mathcal{I}_{\mathcal{D}_1} = \{\langle s, q\rangle\}$
- $\mathcal{I}_{\mathcal{D}_2} = \{\langle br, ar\rangle\}$

From $\mathcal{I}_{\mathcal{D}_1}$, we see that the monitor does not depend on the order between s and q. From $\mathcal{I}_{\mathcal{D}_2}$, we see that the monitor depends on the order between writes themselves, reads, and writes and reads.

[7] In the technical report [68], we extract $\mathcal{I}_\mathcal{D}$ from 55 event-based property specifications from [1,24] written as Quantified Regular Expressions (QREs).

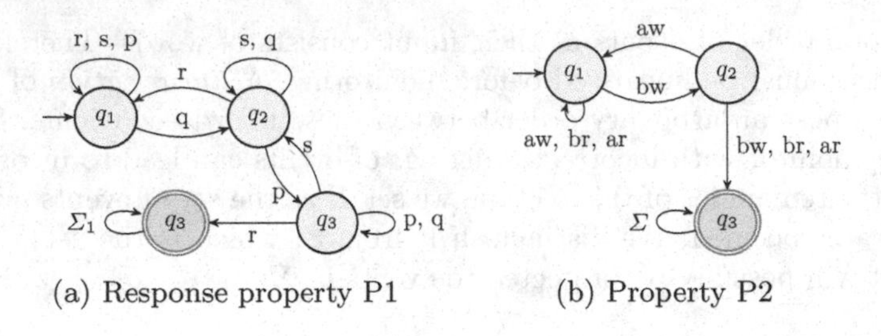

(a) Response property P1 (b) Property P2

Fig. 3. Automata of Bad Prefixes.

6.2 Trace Monitorability of Concurrent Executions

We first define the notion of necessary order for a concurrent trace, which indicates whether the trace has the needed order based on the dependence relation.

Definition 6 (Trace Necessary Order). *We say that a concurrent trace* $t = (\mathcal{E}, \xrightarrow{\text{tr}})$ *has the necessary orderings w.r.t. a causal dependence relation* $\mathcal{D}$*, noted* $\text{tno}(t, \mathcal{D})$ *when:* $\forall\, e, e' \in \mathcal{E} : \langle e, e' \rangle \in \xrightarrow{\text{tr}} \;\vee\; \langle label(e), label(e') \rangle \notin \mathcal{D}$

Example 3 (Trace Necessary Order). Back to property **P2** from Example 2 and the traces depicted in Fig. 2. Trace t_3 is ordered enough for monitoring the property, whereas t_2, a trace collected considering the thread order only, does not capture order between reads and writes. As such we have:
$\text{tno}(t_3, \mathcal{D}_2) = \top$ $\qquad$ $\text{tno}(t_2, \mathcal{D}_2) = \bot$

Let us recall the notion of monitorability from [13,41]. A property φ is *monitorable*, denoted by $\text{Mon}(\varphi)$, if every prefix of every trace has a finite extension that allows its monitor to reach a verdict, be it positive or negative. Monitoring with unsound traces leads to unsound verdicts. We redefine monitorability for concurrent programs by adding necessary conditions on the traces.

Definition 7 (Trace Monitorability of Concurrent Executions). *Given a property* φ *with its dependency relation* $\mathcal{D}$*, and a trace* t *collected from a concurrent execution* e*. Property* φ *is monitorable with* t*, noted* **t-Mon**(φ) *when* $\text{Mon}(\varphi) \wedge \text{snd}(e, t) \wedge \text{tno}(t, \mathcal{D})$.

First, the property φ should be monitorable in the classical sense, $\text{Mon}(\varphi)$, or else we will not reach a verdict. Second, the trace should be sound, $\text{snd}(e, t)$, or else we will have an unsound verdict. Third, the trace should have all the ordering information needed by the property as per its dependency relation, $\text{tno}(t, \mathcal{D})$, or else a linearization would produce an unsound trace. The above indicates that a concurrent trace does not need to contain all the ordering information between events and that the notion of faithfulness can be relaxed when monitoring. Now, if there is missing information in a sound and faithful concurrent trace, this means that the execution itself does not contain the needed causality for monitoring the property. As such, the user is warned about the missing order to address the problem, and a tradeoff is presented between concurrency and monitorability. On one hand, they can synchronize the concurrent actions in the program to have them ordered, or they can force linearization of unordered actions via instrumentation, as discussed in Sect. 2. On the other hand, they can leave the actions executing concurrently and afford inconsistent verdicts.

7 Experimentation and Evaluation

We implemented our work in the tool FACTS (**Fa**ithful and Sound **C**oncurrent **T**races) for Java programs[8]. FACTS consists of three modules; the *instrumentation* module allows the user to specify the instrumentation logic and is built on top of BISM instrumentation framework [69]. The *trace reordering* module implements the vector clock algorithm (Sect. 5) which is responsible for timestamping the events. The *monitorability checker* module, takes as input the causal dependence relation $\mathcal{D}$. After receiving new events, it checks if the collected concurrent trace is monitorable as discussed in Sect. 6.

Experimental Setup. We pick for our evaluation real-world Java applications from Renaissance [59] and DaCapo Benchmarks [16], and synthetic programs from [35]. We monitor properties that can be expressed with total order formalisms to show how concurrent traces can help existing monitoring approaches adapt to concurrent programs. We compare three concurrent trace collection approaches: with **FACTS** in both *asynchronous* and *synchronous* modes, and with Algorithm **A** from [63] (which cannot be run in asynchronous mode)[9]. We instrument the programs to collect the property-related events and the synchronization actions such as thread operations, synchronized blocks, and methods, locks, reads and writes to shared variables, and spawning actors [6] in Akka [2].

Benchmarks. We monitor the program **akka-uct** for a response property stating that between the submission and the execution of a task with normal priority (events q and r resp.) if an urgent task is submitted (event p) it should execute (event s) in between q and r. The property is identical to **P1** from Example 2. For the independence relation, we have $\mathcal{I}_D = \{\langle s, q\rangle\}$. We extract it beforehand using Algorithm 2 from [68] and pass it to FACTS as part of the specification. We monitor program **future-genetic** to check whether dependent tasks execute in parallel. The property is similar to **P2** from Example 2 without q_3 and the *read* events. For the Dacapo benchmarks **avrora** and **fop**, we target the classical type-state properties [70]; the SafeIterator and HasNext properties. For both properties, $\mathcal{I}_D$ is empty and all events are expected to be ordered. We also

[8] Implementation details can be found in the technical report [68].

[9] We also collect linear traces as collected by Java-MOP [21] and show the number of ordering corrections made with concurrent traces, details in [68].

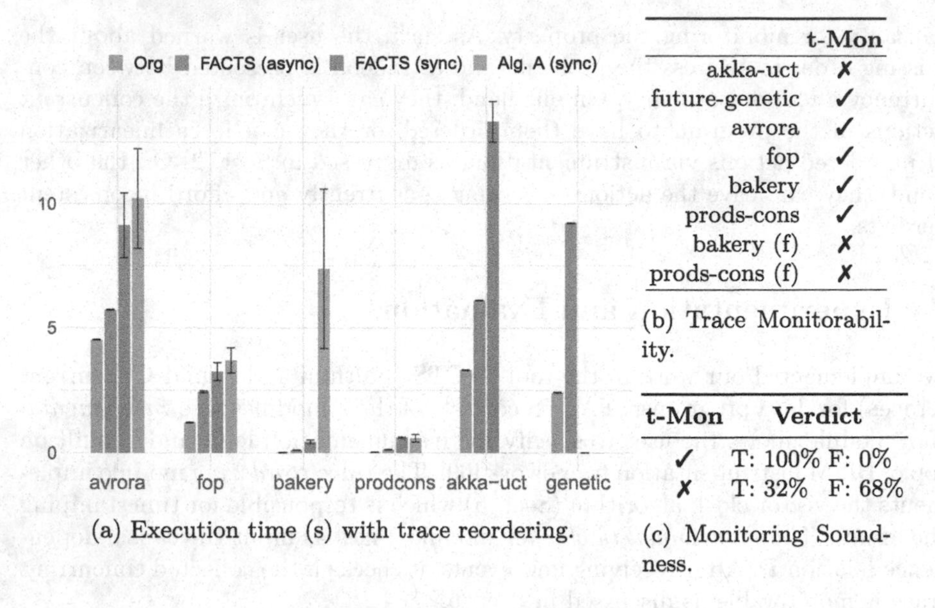

(a) Execution time (s) with trace reordering.

	t-Mon
akka-uct	✗
future-genetic	✓
avrora	✓
fop	✓
bakery	✓
prods-cons	✓
bakery (f)	✗
prods-cons (f)	✗

(b) Trace Monitorability.

t-Mon	Verdict
✓	T: 100% F: 0%
✗	T: 32% F: 68%

(c) Monitoring Soundness.

Fig. 4. Experimentation Results.

monitor the Bakery locking algorithm [43], **bakery**, and a classic producer-consumer program, **prods-cons**. For both, we monitor a property similar to **P2** from Example 2, as such, a similar $\mathcal{I}_D$.

Results. Figure 4a reports the mean execution time for 20 runs of the benchmarked programs. We first note that FACTS is capable of producing sound and faithful traces from all benchmarks, with no marginal cases (i.e., conflicting writes) reported by our vector clock algorithm. This holds even for the *bakery* benchmark, which relies solely on reads and writes for the synchronization of multiple threads. Secondly, running our algorithm in asynchronous mode, FACTS (async), interferes minimally with the program as it incurs a considerably low overhead in most of the benchmarks, relieving the program from blocking when timestamping events and executing the monitors. Thirdly, for the comparable algorithms, FACTS (sync) performs better than Algorithm A in most of the benchmarks. We fairly believe that our algorithm interferes less with the programs as it imposes finer-grained synchronization than Algorithm A. This is highlighted with the bakery algorithm which synchronizes only using shared variables. Algorithm A requires the update of vector clocks associated with a read or write to be atomic through synchronization (as in Sect. 2), while our algorithm does not. This causes the threads to spin more as more contention is added with Algorithm A. Fourthly, for **future-genetic** and **akka-uct**, these programs use parallel tasks and message-passing (resp.) for managing concurrency. We can see how capturing concurrent traces synchronously from them interferes severely with their behavior. Algorithm A and FACTS (sync) timed out with **future-genetic**, while for **akka-uct**, Algorithm A is not intended to handle message passing. Monitoring programs that utilize concurrency primitives with higher levels of abstraction will require improved adaptation in the future. For the time being, it is preferable to observe and monitor such programs asynchronously.

Figure 4b reports on the monitorability of the collected traces. As per our approach, the monitor is warned when **t-Mon** is false, as it may produce unsound verdicts. To test the effectiveness of our approach when used with existing monitoring tools, we introduce two buggy implementations, namely **prods-cons (f)** and **bakery (f)**, where synchronization between threads is faulty, and events can execute without acquiring locks. These programs are intended to have unordered events in the executions (and consequently, the trace), resulting in the trace not having the necessary orderings to monitor the above properties. For each benchmark, we run both a correct implementation where **t-Mon** $= \top$, and a faulty one where **t-Mon** $= \bot$, and monitor them using Java-MOP [21]. We report the results of 100 executions in Fig. 4c. Our findings reveal that monitoring with **t-Mon** $= \bot$ yields inconsistent results, with some executions producing a *true* verdict (32%) while others yield a *false* verdict (68%). This inconsistency highlights the need for our approach, as FACTS produces warnings in all such executions, ensuring sound verdicts. In contrast, verdicts for correct programs are consistent because the execution of events is linearized, and the traces have the necessary orderings.

8 Related Work

We focus on *property-based* dynamic verification techniques for concurrent programs that rely on traces. More specifically, on techniques developed for monitoring behavioral properties expressed in total order formalisms and refer to [15] for a detailed survey. These techniques typically analyze a trace to either *detect* or *predict* violations.

Detection techniques reason about single runs of a program. We mention runtime monitoring tools, namely Java-MOP [21], Tracematches [8,17], MarQ [60], and LARVA [23] chosen from the RV competitions [11,28,61]. These tools allow different specification formalisms such as finite-state machines, extended regular expressions, context-free grammars, past-time linear temporal logic, and Quantified Event Automata (QEA) [10]. Detection techniques do not establish causal orderings between events and rely on trace collection approaches (discussed in Sect. 2) to order the collected events. We have shown in this paper how this can produce unsound traces leading to unsound and inconsistent monitoring. These tools can benefit from concurrent traces to guarantee the soundness of their verdicts. EnforceMOP [49] for instance, can be used to detect and enforce properties (deadlocks as well). It controls the runtime scheduler and blocks threads that might cause a property violation, sometimes leading to a deadlock. It requires forced atomicity as the scheduler needs to decide at each step if the execution on some thread continues or not. In [9,19], the authors present a monitoring framework for actor-based systems. The tool detectEr monitors Erlang applications using traces collected using the native logging functionality. Our approach targets generic concurrency primitives and can also be used with actor-based systems.

Predictive techniques reason about all feasible interleavings from a recorded trace of a single execution. Their challenge is to construct sound and maximal causal models [65] that allow exploring flexibly all feasible interleavings. In [63], the authors present an instrumentation algorithm based on vector clocks, used in [22, 38, 64] for generating the partial order from a running program. The algorithm maintains one vector clock for each thread and two for each shared variable. It executes synchronously with the executing program and is protected using *synchronized* blocks to force an overall sequentially consistent [42] execution. Vector clock algorithms typically require synchronization between the advice, program actions, and algorithm's processing to avoid data races [18]. Our algorithm can run synchronously or asynchronously with the program depending on the monitoring scenario. As far as we know, it is unique in the context of online monitoring in establishing order off the critical path without the need to block the execution to process. In [32, 38] the work targets type-state errors. jPredictor [22] for instance, uses sliced causality [20] to prune the partial order such that only relevant synchronization actions are kept. The tool is demonstrated on atomicity violations and data races; however, we are not aware of an application in the context of generic behavioral properties. In [29], the authors present ExceptioNULL that target null-pointer exceptions. Violations and causality are represented as constraints over actions, and the feasibility of violations is explored via an SMT constraint solver. GPredict [36], for instance, targets generic concurrency properties. It allows the user to express properties with regular expressions and provides explicit concurrency idioms such as atomic and parallel regions. It establishes order by collecting thread-local traces and also producing constraints over actions. In addition to being incomplete due to the possibility of not getting results from the constraint solver, the analysis from these tools might also miss some order relations between events resulting in false positives. Of course, none of the presented predictive techniques are complete, i.e., can produce all possible feasible interleavings that lead to violations, due to the impossibility of constructing a complete causal model of the program. Furthermore, these techniques reason on sequentially consistent execution models [35], restricting the space of possible interleavings of programs. The idea is that if a property is violated in a sequential consistency, then it will surely be violated in a more relaxed execution model. Our work focuses on providing concurrent traces for online detection techniques, and we have yet to explore their applicability in predictive contexts. Unfortunately, many tools from the mentioned approaches [22, 36, 38, 64] are not available. Apart from tools dedicated to data races and atomicity violation detection, we found no available tools targeting general behavioral properties to compare with; tools that can establish causal order and instrument various custom regular actions from the program for monitoring.

9 Conclusion and Future Work

Collecting concurrent traces for runtime monitoring opens the path to interesting problems. Firstly, monitoring techniques can be revisited and extended to take into account the partial order. Tools relying on total order in traces can use concurrent traces to check if a trace has the needed causality and, if not, produce warnings. Moreover, specifications (and formalisms) that can match traces obtained from FACTS can be elaborated to extend the expressiveness of monitoring to check concurrency-related behavioral properties. For example, our approach is applicable for OpenMP runtimes; it can be used to verify the correctness of the scheduling algorithms of tasks with data dependencies (e.g., [71]). We can verify that the runtime never schedules two dependent tasks in parallel. Secondly, it is now possible to define and quantify optimizations for capturing sound and faithful traces. How to obtain optimal faithfulness with instrumentation is still an interesting challenge. Thirdly, for scenarios where the execution does not have the causality needed between events, the user can be given facilities to preserve some safety property by enforcing the order. Linearization of concurrent events may be achieved on the fly by JVM hot swapping.

Finally, we will reduce overheads induced by capturing synchronization actions. We aim to include plugins to FACTS to instrument concurrency primitives with higher-level abstraction such as fork-join [45] and software transactional memory [34]. Tailoring instrumentation to these frameworks reduces the number of collected synchronization actions. It requires assumptions about the correctness of these frameworks. These assumptions can be checked using static and dynamic analysis.

References

1. Patterns in property specifications for finite-state verification home page. https://matthewbdwyer.github.io/psp/patterns.html
2. Akka documentation. http://akka.io/docs/ (2022)
3. JSR-133: Java Memory Model and Thread Specification Revision (2004). http://jcp.org/jsr/detail/133.jsp
4. Adve, S.V., Gharachorloo, K.: Shared memory consistency models: a tutorial. Computer **29**(12), 66–76 (1996). https://doi.org/10.1109/2.546611
5. Agarwal, A., Garg, V.K.: Efficient dependency tracking for relevant events in shared-memory systems. In: Proceedings of the Twenty-Fourth Annual ACM Symposium on Principles of Distributed Computing, pp. 19–28. PODC 2005, Association for Computing Machinery, New York, NY, USA (2005). https://doi.org/10.1145/1073814.1073818
6. Agha, G.: Actors: A Model of Concurrent Computation in Distributed Systems. MIT Press, Cambridge, MA, USA (1986)
7. Ahamad, M., Neiger, G., Burns, J.E., Kohli, P., Hutto, P.W.: Causal memory: definitions, implementation, and programming. Distrib. Comput. **9**(1), 37–49 (1995). https://doi.org/10.1007/BF01784241

8. Allan, C., et al.: Adding trace matching with free variables to AspectJ. In: Proceedings of the 20th Annual ACM SIGPLAN Conference on Object-oriented Programming, Systems, Languages, and Applications, pp. 345–364. OOPSLA 2005, ACM (2005). https://doi.org/10.1145/1094811.1094839
9. Attard, D.P., Cassar, I., Francalanza, A., Aceto, L., Ingólfsdóttir, A.: A runtime monitoring tool for actor-based systems (2022). https://www.um.edu.mt/library/oar/handle/123456789/23062
10. Barringer, H., Falcone, Y., Havelund, K., Reger, G., Rydeheard, D.E.: Quantified event automata: towards expressive and efficient runtime monitors. In: FM 2012: Formal Methods - 18th International Symposium, Paris, France, 27–31 August 2012. Proceedings, pp. 68–84 (2012). https://doi.org/10.1007/978-3-642-32759-9_9
11. Bartocci, E., et al.: First international Competition on Runtime Verification: rules, benchmarks, tools, and final results of CRV 2014. Int. J. Softw. Tools Technol. Transfer **21**(1), 31–70 (2017). https://doi.org/10.1007/s10009-017-0454-5
12. Bartocci, E., Falcone, Y., Francalanza, A., Reger, G.: Introduction to runtime verification. In: Bartocci, E., Falcone, Y. (eds.) Lectures on Runtime Verification. LNCS, vol. 10457, pp. 1–33. Springer, Cham (2018). https://doi.org/10.1007/978-3-319-75632-5_1
13. Bauer, A., Leucker, M., Schallhart, C.: Runtime verification for LTL and TLTL. ACM Trans. Softw. Eng. Methodol. **20**(4), 1–64 (2011). https://doi.org/10.1145/2000799.2000800
14. Bensalem, S., Havelund, K.: Dynamic deadlock analysis of multi-threaded programs. In: Ur, S., Bin, E., Wolfsthal, Y. (eds.) HVC 2005. LNCS, vol. 3875, pp. 208–223. Springer, Heidelberg (2006). https://doi.org/10.1007/11678779_15
15. Bianchi, F.A., Margara, A., Pezzè, M.: A survey of recent trends in testing concurrent software systems. IEEE Trans. Software Eng. **44**(8), 747–783 (2018). https://doi.org/10.1109/TSE.2017.2707089
16. Blackburn, S.M., et al.: The DaCapo benchmarks: java benchmarking development and analysis. In: Proceedings of the 21st Annual ACM SIGPLAN Conference on Object-oriented Programming Systems, Languages, and Applications, pp. 169–190. OOPSLA 2006, ACM (2006). https://doi.org/10.1145/1167473.1167488
17. Bodden, E., Hendren, L., Lam, P., Lhoták, O., Naeem, N.A.: Collaborative Runtime Verification with Tracematches. J. Log. Comput. **20**(3), 707–723 (2010). https://doi.org/10.1093/logcom/exn077
18. Cain, H.W., Lipasti, M.H.: Verifying sequential consistency using vector clocks. In: Proceedings of the Fourteenth Annual ACM Symposium on Parallel Algorithms and Architectures, pp. 153–154. SPAA 2002, Association for Computing Machinery, New York, NY, USA (2002). https://doi.org/10.1145/564870.564897
19. Cassar, I., Francalanza, A.: On implementing a monitor-oriented programming framework for actor systems. In: Integrated Formal Methods - 12th International Conference, IFM 2016, Reykjavik, Iceland, 1–5 June 2016, Proceedings, pp. 176–192 (2016). https://doi.org/10.1007/978-3-319-33693-0_12
20. Chen, F., Roşu, G.: Parametric and sliced causality. In: Damm, W., Hermanns, H. (eds.) CAV 2007. LNCS, vol. 4590, pp. 240–253. Springer, Heidelberg (2007). https://doi.org/10.1007/978-3-540-73368-3_27
21. Chen, F., Roşu, G.: Java-MOP: a monitoring oriented programming environment for java. In: Halbwachs, N., Zuck, L.D. (eds.) TACAS 2005. LNCS, vol. 3440, pp. 546–550. Springer, Heidelberg (2005). https://doi.org/10.1007/978-3-540-31980-1_36

22. Chen, F., Serbanuta, T.F., Rosu, G.: jPredictor: a predictive runtime analysis tool for java. In: Proceedings of the 30th International Conference on Software Engineering, pp. 221–230. ICSE 2008, Association for Computing Machinery, New York, NY, USA (2008). https://doi.org/10.1145/1368088.1368119
23. Colombo, C., Pace, G.J., Schneider, G.: LARVA – Safer monitoring of real-time java programs (tool paper). In: Hung, D.V., Krishnan, P. (eds.) Seventh IEEE International Conference on Software Engineering and Formal Methods, SEFM 2009, Hanoi, Vietnam, 23–27 November 2009, pp. 33–37. IEEE Computer Society (2009). https://doi.org/10.1109/SEFM.2009.13
24. Dwyer, M.B., Avrunin, G.S., Corbett, J.C.: Patterns in property specifications for finite-state verification. In: Proceedings of the 21st International Conference on Software Engineering, pp. 411–420. ICSE 1999, Association for Computing Machinery, New York, NY, USA (1999). https://doi.org/10.1145/302405.302672
25. El-Hokayem, A., Falcone, Y.: Can we monitor all multithreaded programs? In: Colombo, C., Leucker, M. (eds.) RV 2018. LNCS, vol. 11237, pp. 64–89. Springer, Cham (2018). https://doi.org/10.1007/978-3-030-03769-7_6
26. Falcone, Y., Fernandez, J., Mounier, L.: What can you verify and enforce at runtime? Int. J. Softw. Tools Technol. Transf. **14**(3), 349–382 (2012). https://doi.org/10.1007/s10009-011-0196-8
27. Falcone, Y., Havelund, K., Reger, G.: A Tutorial on Runtime Verification. In: Engineering Dependable Software Systems, pp. 141–175. IOS Press (2013)
28. Falcone, Y., Ničković, D., Reger, G., Thoma, D.: Second international competition on runtime verification. In: Bartocci, E., Majumdar, R. (eds.) RV 2015. LNCS, vol. 9333, pp. 405–422. Springer, Cham (2015). https://doi.org/10.1007/978-3-319-23820-3_27
29. Farzan, A., Madhusudan, P., Razavi, N., Sorrentino, F.: Predicting null-pointer dereferences in concurrent programs. In: Proceedings of the ACM SIGSOFT 20th International Symposium on the Foundations of Software Engineering. FSE 2012, Association for Computing Machinery, New York, NY, USA (2012). https://doi.org/10.1145/2393596.2393651
30. Flanagan, C., Freund, S.N.: Atomizer: A dynamic atomicity checker for multithreaded programs. SIGPLAN Not. **39**(1), 256–267 (2004). https://doi.org/10.1145/982962.964023
31. Flanagan, C., Freund, S.N.: FastTrack: Efficient and precise dynamic race detection. In: Proceedings of the 30th ACM SIGPLAN Conference on Programming Language Design and Implementation, pp. 121–133. PLDI 2009, Association for Computing Machinery, New York, NY, USA (2009). https://doi.org/10.1145/1542476.1542490
32. Gao, Q., Zhang, W., Chen, Z., Zheng, M., Qin, F.: 2ndstrike: Toward manifesting hidden concurrency typestate bugs. SIGPLAN Not. **46**(3), 239–250 (2011). https://doi.org/10.1145/1961296.1950394
33. Gastin, P., Kuske, D.: Uniform satisfiability problem for local temporal logics over Mazurkiewicz traces. Inf. Comput. **208**(7), 797–816 (2010). https://doi.org/10.1016/j.ic.2009.12.003
34. Harris, T., Marlow, S., Peyton-Jones, S., Herlihy, M.: Composable memory transactions. In: Proceedings of the Tenth ACM SIGPLAN Symposium on Principles and Practice of Parallel Programming, pp. 48–60. PPoPP 2005, Association for Computing Machinery, New York, NY, USA (2005). https://doi.org/10.1145/1065944.1065952
35. Herlihy, M., Shavit, N.: The Art of Multiprocessor Programming, 1st edn. Revised Reprint. Morgan Kaufmann Publishers Inc., San Francisco, CA, USA (2012)

36. Huang, J., Luo, Q., Rosu, G.: Gpredict: generic predictive concurrency analysis. In: 37th IEEE/ACM International Conference on Software Engineering, ICSE 2015, vol. 1, pp. 847–857 (2015). https://doi.org/10.1109/ICSE.2015.96
37. Huang, J., Meredith, P.O., Rosu, G.: Maximal sound predictive race detection with control flow abstraction. In: Proceedings of the 35th ACM SIGPLAN Conference on Programming Language Design and Implementation. p. 337–348. PLDI 2014, Association for Computing Machinery, New York, NY, USA (2014). https://doi.org/10.1145/2594291.2594315
38. Joshi, P., Sen, K.: Predictive typestate checking of multithreaded java programs. In: Proceedings of the 2008 23rd IEEE/ACM International Conference on Automated Software Engineering, pp. 288–296. ASE 2008, IEEE Computer Society, USA (2008). https://doi.org/10.1109/ASE.2008.39
39. Joshi, Y., Tchamgoue, G.M., Fischmeister, S.: Runtime verification of ITL on lossy traces. In: Proceedings of the Symposium on Applied Computing (2017)
40. Kiczales, G., Hilsdale, E., Hugunin, J., Kersten, M., Palm, J., Griswold, W.G.: Getting started with AspectJ. Commun. ACM **44**(10), 59–65 (2001)
41. Kupferman, O., Y. Vardi, M.: Model checking of safety properties. Form. Methods Syst. Des. **19**(3), 291–314 (2001). https://doi.org/10.1023/A:1011254632723
42. Lamport: How to make a multiprocessor computer that correctly executes multiprocess programs. IEEE Trans. Comput. **C-28**(9), 690–691 (1979). https://doi.org/10.1109/TC.1979.1675439
43. Lamport, L.: A new solution of Dijkstra's concurrent programming problem. Commun. ACM **17**(8), 453–455 (1974). https://doi.org/10.1145/361082.361093
44. Lamport, L.: Time, clocks, and the ordering of events in a distributed system. Commun. ACM **21**(7), 558–565 (1978). https://doi.org/10.1145/359545.359563
45. Lea, D.: A java fork/join framework. In: Proceedings of the ACM 2000 Java Grande Conference, San Francisco, CA, USA, 3–5 June 2000, pp. 36–43 (2000). https://doi.org/10.1145/337449.337465
46. Lea, D., Bloch, J., Midkiff, S., Holmes, D., Bowbeer, J., Peierls, T.: JSR-166: Concurrency utilities. https://www.jcp.org/en/jsr/detail?id=166
47. Leucker, M., Schallhart, C.: A brief account of runtime verification. J. Logic Algebraic Program. **78**(5), 293–303 (2009). https://doi.org/10.1016/j.jlap.2008.08.004
48. Lodaya, K., Weil, P.: Rationality in algebras with a series operation. Inf. Comput. **171**(2), 269–293 (2001). https://doi.org/10.1006/inco.2001.3077
49. Luo, Q., Rosu, G.: EnforceMOP: a runtime property enforcement system for multithreaded programs. In: Proceedings of International Symposium in Software Testing and Analysis (ISSTA2013), pp. 156–166. ACM (2013). http://dl.acm.org/citation.cfm?doid=2483760.2483766
50. Manna, Z., Pnueli, A.: A hierarchy of temporal properties (invited paper, 1989). In: Proceedings of the Ninth Annual ACM Symposium on Principles of Distributed Computing, pp. 377–410. PODC 1990, Association for Computing Machinery, New York, NY, USA (1990). https://doi.org/10.1145/93385.93442
51. Manna, Z., Pnueli, A.: Temporal Verification of Reactive Systems: Safety. Springer-Verlag, New York, Inc (1995). https://doi.org/10.1007/978-1-4612-4222-2
52. Manson, J., Pugh, W., Adve, S.V.: The java memory model. In: Proceedings of the 32nd ACM SIGPLAN-SIGACT Symposium on Principles of Programming Languages, pp. 378–391. POPL 2005, ACM (2005). https://doi.org/10.1145/1040305.1040336
53. Mathur, U., Viswanathan, M.: Atomicity checking in linear time using vector clocks, pp. 183–199. Association for Computing Machinery, New York, NY, USA (2020). https://doi.org/10.1145/3373376.3378475

54. Mattern, F.: Virtual time and global states of distributed systems. In: Parallel and Distributed Algorithms, pp. 215–226. North-Holland (1988)
55. Mazurkiewicz, A.: Trace theory. In: Brauer, W., Reisig, W., Rozenberg, G. (eds.) ACPN 1986. LNCS, vol. 255, pp. 278–324. Springer, Heidelberg (1987). https://doi.org/10.1007/3-540-17906-2_30
56. Meenakshi, B., Ramanujam, R.: Reasoning about layered message passing systems. Comput. Lang. Syst. Struct. **30**(3–4), 171–206 (2004). https://doi.org/10.1016/j.cl.2004.02.003
57. Nielsen, M., Plotkin, G.D., Winskel, G.: Petri nets, event structures and domains, part I. Theor. Comput. Sci. **13**, 85–108 (1981). https://doi.org/10.1016/0304-3975(81)90112-2
58. Peled, D., Havelund, K.: Refining the safety–liveness classification of temporal properties according to monitorability. In: Margaria, T., Graf, S., Larsen, K.G. (eds.) Models, Mindsets, Meta: The What, the How, and the Why Not? LNCS, vol. 11200, pp. 218–234. Springer, Cham (2019). https://doi.org/10.1007/978-3-030-22348-9_14
59. Prokopec, A., et al.: Renaissance: a modern benchmark suite for parallel applications on the JVM. In: Proceedings Companion of the 2019 ACM SIGPLAN International Conference on Systems, Programming, Languages, and Applications: Software for Humanity, pp. 11–12. SPLASH Companion 2019, Association for Computing Machinery, New York, NY, USA (2019). https://doi.org/10.1145/3359061.3362778
60. Reger, G., Cruz, H.C., Rydeheard, D.: MarQ: monitoring at runtime with QEA. In: Baier, C., Tinelli, C. (eds.) TACAS 2015. LNCS, vol. 9035, pp. 596–610. Springer, Heidelberg (2015). https://doi.org/10.1007/978-3-662-46681-0_55
61. Reger, G., Hallé, S., Falcone, Y.: Third international competition on runtime verification. In: Falcone, Y., Sánchez, C. (eds.) RV 2016. LNCS, vol. 10012, pp. 21–37. Springer, Cham (2016). https://doi.org/10.1007/978-3-319-46982-9_3
62. Roemer, J., Genç, K., Bond, M.D.: SmartTrack: efficient predictive race detection. In: Proceedings of the 41st ACM SIGPLAN Conference on Programming Language Design and Implementation, pp. 747–762. PLDI 2020, Association for Computing Machinery, New York, NY, USA (2020). https://doi.org/10.1145/3385412.3385993
63. Rosu, G., Sen, K.: An instrumentation technique for online analysis of multithreaded programs. In: 18th International Parallel and Distributed Processing Symposium, 2004. Proceedings, pp. 268- (2004). https://doi.org/10.1109/IPDPS.2004.1303344
64. Sen, K., Rosu, G., Agha, G.: Runtime safety analysis of multithreaded programs. SIGSOFT Softw. Eng. Notes **28**(5), 337–346 (2003). https://doi.org/10.1145/949952.940116
65. Serbanuta, T., Chen, F., Rosu, G.: Maximal causal models for sequentially consistent systems. In: Runtime Verification, Third International Conference, RV 2012, Istanbul, Turkey, 25–28 September 2012, Revised Selected Papers, pp. 136–150 (2012). https://doi.org/10.1007/978-3-642-35632-2_16
66. Shin, K., Ramanathan, P.: Real-time computing: a new discipline of computer science and engineering. Proc. IEEE **82**(1), 6–24 (1994). https://doi.org/10.1109/5.259423
67. Soueidi, C., El-Hokayem, A., Falcone, Y.: Opportunistic monitoring of multithreaded programs. In: Lambers, L., Uchitel, S. (eds.) FASE 2023. LNCS, vol. 13991, pp. 173–194. Springer, Cham (2023). https://doi.org/10.1007/978-3-031-30826-0_10

68. Soueidi, C., Falcone, Y.: Sound concurrent traces for online monitoring technical report (2023). https://doi.org/10.6084/m9.figshare.21940205
69. Soueidi, C., Kassem, A., Falcone, Y.: BISM: bytecode-level instrumentation for software monitoring. In: Deshmukh, J., Ničković, D. (eds.) RV 2020. LNCS, vol. 12399, pp. 323–335. Springer, Cham (2020). https://doi.org/10.1007/978-3-030-60508-7_18
70. Strom, R.E., Yemini, S.: Typestate: A programming language concept for enhancing software reliability. IEEE Trans. Soft. Eng. **12**(1), 157–171 (1986). https://doi.org/10.1109/TSE.1986.6312929
71. Virouleau, P., Broquedis, F., Gautier, T., Rastello, F.: Using data dependencies to improve task-based scheduling strategies on NUMA architectures. In: Dutot, P.-F., Trystram, D. (eds.) Euro-Par 2016. LNCS, vol. 9833, pp. 531–544. Springer, Cham (2016). https://doi.org/10.1007/978-3-319-43659-3_39
72. Wang, L., Stoller, S.: Runtime analysis of atomicity for multithreaded programs. IEEE Trans. Software Eng. **32**(2), 93–110 (2006). https://doi.org/10.1109/TSE.2006.1599419

Testing

Efficient Trace Generation for Rare-Event Analysis in Chemical Reaction Networks

Bryant Israelsen(✉), Landon Taylor, and Zhen Zhang

Utah State University, Logan, UT, USA
{bryant.israelsen,landon.jeffrey.taylor,zhen.zhang}@usu.edu

Abstract. Rare events are known to potentially cause pathological behavior in biochemical reaction systems. It is important to understand the cause. However, rare events are challenging to analyze due to their extremely low observability. This paper presents a fully automated approach that rapidly generates a large number of execution traces guaranteed to reach user-specified rare-event states for Chemical Reaction Network (CRN) models. It is enabled by a unique combination of a multi-layered and service-oriented CRN formal modeling approach, a dependency graph method to aid the shortest rare-event trace generation, and randomized compositional testing. The resulting prototype tool shows marked improvement over stochastic simulation and probabilistic model checking and it offers insights into a CRN.

Keywords: Compositional testing · rare events · dependency graph

1 Introduction

As a formalism for modeling chemical kinetics, *Chemical Reaction Networks* (CRNs) are widely used for modeling biochemical reaction systems [6], genetic regulatory networks [29], and molecular programming [38]. Many biochemical systems are intrinsically stochastic, including processes in gene and protein expressions. Essentially, their constituent chemical reactions are often simultaneously enabled to occur in parallel with different probabilities. Moreover, their noisy operating environment can introduce unexpected behavior. Rare events in these systems are often of significant relevance, because they represent extreme infrequent occurrence of undesirable behavior that may lead to pathological effects. Therefore, obtaining provable reliability guarantees is a must for CRNs. *Probabilistic model checking* (PMC) can provide such quantitative guarantees and allows *in silico* analysis for detecting and quantifying rare errors. However, PMC approaches are challenged by the need for enumerating a model's large or even infinite state space to gather a sufficient number of rare-event traces in order to provide accurate probability verification. This task is typically computationally intractable. Further, for probabilistic analysis, it is often necessary to gather many traces that reach the rare-event states of interest. Generating only a small number of them is often insufficient to give an accurate estimate.

G. Caltais and C. Schilling (Eds.): SPIN 2023, LNCS 13872, pp. 83–102, 2023.
https://doi.org/10.1007/978-3-031-32157-3_5

This paper presents a fully automated approach that rapidly generates a large number of execution traces guaranteed to satisfy a user-specified rare-event property for a CRN model. These traces are used to compute a lower probability bound for the rare-event property. We first propose a novel multi-layered, service-oriented, and modular CRN modeling approach using the IVy modeling language [27]. It offers flexibility in customizing both the reaction execution frequency and the length of traces. We then propose a dependency graph method to guide the shortest trace generation with unique finite prefixes through compositional testing. These traces are guaranteed to reach the specified rare event and are collected to compute the rare-event's lower probability bound. The dependency graph also effectively proves unreachability of a given rare event, leading to considerable savings in performance. We implemented these methods in a prototype tool, *Random Assume Guarantee Testing Induced Model Executions for Reachability* (RAGTIMER), and found preliminary results to be promising. The proposed rare-event trace enumeration technique can potentially be integrated with many formal and semi-formal rare-event analysis methods and the generated traces can provide detailed debugging information for understanding reachability of rare events. We believe that this unique combination of the presented methods has not been proposed elsewhere, and is an effective alternative to rare-event simulation approaches for biochemical reaction networks.

2 Preliminaries

Chemical Reaction Networks (CRNs). Under the *Stochastic Chemical Kinetic* (SCK) model assumption, the time-evolution of a CRN is governed by the Chemical Master Equation. Formally, a CRN is a tuple $\mathcal{M}$ composed of n chemical species $\mathcal{X} = \{\mathcal{X}_1, \ldots, \mathcal{X}_n\}$, m reactions $\mathcal{R} = \{\mathcal{R}_1, \ldots, \mathcal{R}_m\}$, and an initial state representing each species' molecule count $s_0 : \mathcal{X}^n \rightarrow \mathbb{Z}_{\geqslant 0}$. Given a reaction $\mathcal{R}_i$, denote $\mathsf{Reactant}_i \subseteq \mathcal{X}$ as the reactant set and $\mathsf{Product}_i \subseteq \mathcal{X}$ as the product set in $\mathcal{R}_i$. A reaction $\mathcal{R}_i = \langle \alpha_i, v_i \rangle$ includes a *propensity function* $\alpha_i : \mathbb{Z}^n_{\geqslant 0} \rightarrow \mathbb{R}^+$ corresponding to the probability (including 0) for it to occur in a state and the *state change vector* $v_i \in \mathbb{Z}^n$ corresponding to the update in molecule count for each species due to reaction $\mathcal{R}_i$. Under the SCK assumption, each reaction $\mathcal{R}_i$ occurs nearly instantaneously, which practically limits both v_i to the values of $0, \pm 1, \pm 2$, and the size of $\mathsf{Reactant}_i$ to be less than three [29].

CRN Semantics. A CRN under the SCK assumption induces a *Continuous-time Markov Chain* (CTMC), where state change due to a reaction occurs in discrete amounts and the probability of state change is dependent on real-valued time. A CTMC model $\mathcal{C}$ is a tuple $\mathcal{C} = \langle \mathbf{S}, s_0, \mathbf{R}, \mathbf{L} \rangle$ where $\mathbf{S}$ is a finite state set (i.e., *state space*); $s_0 \in \mathbf{S}$ is the sole initial state; $\mathbf{R} : \mathbf{S} \times \mathbf{S} \rightarrow \mathbb{R}_{\geqslant 0}$ is the transition rate matrix; and $\mathbf{L} : \mathbf{S} \rightarrow 2^{AP}$ is a state labeling function with atomic propositions AP. Transition rate $\mathbf{R}(s, s')$ from state s to s' is determined by the propensity of $\mathcal{R}_i$, assuming $\mathcal{R}_i$ is the sole reaction causing this state change. The propensity is the number of possible combinations of reactant molecules:

$\alpha_i(s) = k_i \prod_{\mathcal{X}_j \in \mathsf{Reactant}_i}(s[j])$, where $\mathcal{R}_i$'s *reaction rate constant* is $k_i \in \mathbb{R}^+$. A reaction $\mathcal{R}_i$ is *enabled* to occur in state s if its corresponding propensity function evaluates to a positive value, i.e., $\alpha_i(s) > 0$. Often, multiple reactions are enabled to occur in state s and the corresponding probability for $\mathcal{R}_i$ is $p(s, s') = \frac{\mathbf{R}(s,s')}{\mathbf{E}(s)}$, where the *exit rate* $\mathbf{E}(s) = \sum_{s' \in post(s)} \mathbf{R}(s, s')$ sums up all enabled reaction rates in s. A CTMC model has a non-zero probability of staying in a state and the probability of leaving a state s within time interval $[0, t]$ is $1 - e^{\mathbf{E}(s) \cdot t}$, where t is a non-negative real-valued quantity representing real time.

Time-bounded Reachability Property and Target States. We focus on computing the following non-nested time-bounded transient reachability probability specified in *Continuous Stochastic Logic* (CSL) [3,20]: $\mathsf{P}_{=?}(\Diamond^{[0,T]}\ \Psi)$. It queries the probability of reaching the rare-event Ψ-states within T time units. Let condition Ψ be $\mathcal{X}_\Psi = C_\Psi$, where $C_\Psi \in \mathbb{Z}_{\geqslant 0}$ and $s_0(\mathcal{X}_\Psi) \neq C_\Psi$. That is, a target is an equality condition for exactly one species and it is not initialized to the target value. A state is a *target state* s_Ψ if Ψ evaluates to `true` in s_Ψ, i.e., $s_\Psi \models \Psi$. This work aims at efficiently providing the guaranteed lower probability bound, i.e., $\mathsf{P}_{min}(\Diamond^{[0,T]}\ \Psi)$, where Ψ is $\mathcal{X}_\Psi = C_\Psi$. Note that the user is not required to provide an upper bound for each species, which could induce an infinite-state CTMC. However, the method presented in this paper only generates finite traces where the last state is a target state. Therefore, the resulting CTMC constructed from these traces have a finite state space.

Compositional Testing. A CRN model $\mathcal{M}$ consists of interacting chemical reactions where each executes atomically. $\mathcal{M}$ is a *closed* system, meaning that it does not require any external input, because reactants required by one reaction are provided by other reactions in the same model. These features naturally allow for compositional testing. As detailed in Sect. 6, a CRN model can be represented as a composition of two interacting processes p_1 and p_2, denoted as $p_1 \| p_2$, following the circular assume-guarantee reasoning rules shown below:

$$\frac{\langle\alpha\rangle\ p_1\ \langle\gamma\rangle \quad \langle\gamma\rangle\ p_2\ \langle\alpha\rangle}{\langle\text{true}\rangle\ p_1 \| p_2\ \langle\alpha \wedge \gamma\rangle} \quad (1) \qquad \frac{\langle\alpha\rangle\ p_1\ \langle\alpha\rangle \quad \langle\alpha\rangle\ p_2\ \langle\alpha\rangle}{\langle\text{true}\rangle\ p_1 \| p_2\ \langle\alpha\rangle} \quad (2)$$

The triple $\langle\alpha\rangle p_1 \langle\gamma\rangle$ in Rule (1) can be understood as follows. From the start of process execution, up to step $k-1$, if p_1 satisfies its environment *assumption* α, in the form of Boolean-valued constraints on p_1's input, then the allowed input and output behavior of p_1 determines the *guarantee* γ in the current execution step k. Similarly, process p_2 in the triple $\langle\gamma\rangle p_2 \langle\alpha\rangle$ guarantees α at the present if γ holds in the past. This interpretation avoids the circular definition of this rule by requiring that each process in the composition *only* relies on the correctness of inputs it received in the *past*, but not those to be received in the future, in order for its output to satisfy their respective guarantees. Therefore, as long as neither assumption fails first, neither guarantee can fail first in the composition, and hence $p_1 \| p_2 \models \alpha \wedge \gamma$. For the triple $\langle\alpha\rangle p_1 \langle\alpha\rangle$ in Rule (2), it is interpreted as that p_1 does not cause the global property α to fail. This rule states that if neither process in the composition causes α to fail first, then α always holds.

Predicated on Rules (1) and (2), *compositional testing* [11,25,28] is a semi-formal technique that empirically checks satisfiability of the guarantee γ for each triple $\langle\alpha\rangle p_i\langle\gamma\rangle$ in the composition by *sampling* inputs from those satisfying the assumption α. Generation of inputs typically involves randomization. For Rule (2), testing of the triple $\langle\alpha\rangle p_i\langle\alpha\rangle$ includes both generating only inputs satisfying α and verifying that the outputs of p_i do not fail α. Each process $p_i = (I_i, O_i)$ consists of an input action set I_i and an output action set O_i. Denote a_i^m as action a_m *defined* in process p_i. a_i^m is an *input action* for p_i if a_i^m is called by another process p_j ($j \neq i$), but is an *output action* for p_j. Execution of a_i^m modifies a non-empty set of local variables in p_i, and each variable $v \in V_i$ is bounded by a range R_v. Two processes p_1 and p_2 are compatible for composition if $O_1 \cap O_2 = \emptyset$, and their composition is $p_1 \| p_2 = (I, O)$, where $I = (I_1 \cup I_2)\backslash(O_1 \cup O_2)$ and $O = (O_1 \cup O_2)$.

3 Related Work

Rare-event properties are challenging to analyze due to their extremely low observability. *Statistical model checking* (SMC) techniques (e.g., [24,30,43]) have integrated rare-event methods, including *importance sampling* [14,16,17] and *importance splitting* [18,36,42]. Importance sampling biases simulation by weighting the rare-event probability to increase its observability and then compensates for the loss to yield the unbiased probability [24]. Importance splitting reformulates a rare-event probability as a product of less-rare level-conditional probabilities [15]. For analyzing rare-events in biochemical systems, the *weighted Stochastic Simulation Algorithm* (wSSA) [19] relies on a user-defined biasing scheme to favor reactions leading to observing the rare-event, but is limited by the user's insight in selecting the proper biasing scheme. Extensions of wSSA (e.g., [14,33]) have substantially improved its computational efficiency. Recent algorithms (e.g., [35]) can characterize rare events in terms of system parameters. As an alternative, *weighted ensemble* [2,44] has been configured to sample rare events in CRNs [9,45]. Importance splitting divides a model's state space into contiguous levels ordered in the increasing likelihood of reaching the rare event [23,24,41]. The crux of it is the (possibly manually constructed) importance function, which rewards a simulation trace by spawning multiple copies if it crosses a level closer to the rare event, but terminates it otherwise. In [4], the authors presented an automated compositional importance function derivation technique based on the model structure and the rare-event property. Recently, the extended RESTART with *prolonged retrials* [39,40] importance technique was re-implemented in the SMC engine `modes` [4,5] in the MODEST TOOLSET [12].

Advantages of the Proposed Approach over Rare-event Simulation. First, it is *fully automated* and neither requires expert knowledge of nor poses modeling limitations on the CRN model. Secondly, it is potentially less computationally intensive as it neither requires rare-event biasing computation nor wastes any simulation traces not able to reach the rare event. Lastly, simulation-based approaches provide an *estimate* of the actual rare-event probability, whereas the proposed method provides a guaranteed lower probability bound.

4 Motivating Example

The motivating example is the *modified yeast polarization* model [7], a CRN consisting of seven species reacting through eight reactions:

$$\mathcal{R}_1 : \emptyset \xrightarrow{0.0038} \mathrm{R}, \quad \mathcal{R}_2 : \mathrm{R} \xrightarrow{4.00\times10^{-4}} \emptyset, \quad \mathcal{R}_3 : \mathrm{L}+\mathrm{R} \xrightarrow{0.042} \mathrm{RL}+\mathrm{L},$$
$$\mathcal{R}_4 : \mathrm{RL} \xrightarrow{0.010} \mathrm{R}, \quad \mathcal{R}_5 : \mathrm{RL}+\mathrm{G} \xrightarrow{0.011} \mathrm{G_a}+\mathrm{G_{bg}}, \quad \mathcal{R}_6 : \mathrm{G_a} \xrightarrow{0.100} \mathrm{G_d},$$
$$\mathcal{R}_7 : \mathrm{G_d}+\mathrm{G_{bg}} \xrightarrow{1.05\times10^{3}} \mathrm{G}, \quad \mathcal{R}_8 : \emptyset \xrightarrow{3.21} \mathrm{RL}.$$

All reaction propensities are in molecules per second. The initial molecule count for the following species vector $(R, L, RL, G, G_a, G_{bg}, G_d)$ forms the initial state $s_0 = [50, 2, 0, 50, 0, 0, 0]$. This system was modified from the pheromone induced G-protein cycle in Saccharomyces cerevisia [10] with a constant population of ligand (L = 2) preventing it from reaching equilibrium [34]. The rare event is a measure of an unreasonably rapid build-up of G_{bg}. Thus, the property of interest is the probability that the molecule count for G_{bg} reaching 50 within 20 seconds: $\mathsf{P}_{=?}(\Diamond^{[0,20]}\, G_{bg} = 50)$. The high concurrency nature of this model is evidenced by $\mathcal{R}_1$ and $\mathcal{R}_8$ each being independent of all other reactions and enabled in all states. Additionally, it takes at least 100 reaction executions to reach a target state. These features can easily overwhelm state expansion methods performed by probabilistic model checking tools as discussed in Sect. 10.

5 Method Overview

Figure 1 shows a logical flow of the proposed novel approach for the RAGTIMER tool, where the steps in blue symbolize looping behavior. It first reads in a user-specified CRN $\mathcal{M}$ and a rare-event property of interest. A dependency graph is then generated for the given CRN model and target property reachability is determined, as described in Sect. 7. The dependency graph information is then used to automatically generate the service-oriented layered IVy model (Sect. 6), which is used with compositional testing to generate the desirable shortest traces with unique prefixes as described in Sects. 7 and 8. Stochastic simulation is then performed on each trace to obtain its execution probability, and a summary of these results is returned to the user (Sect. 9).

RAGTIMER significantly differs from statistical model checking techniques that *estimate* the rare-event probability by biasing events leading to the rare-event during stochastic simulation (e.g., importance sampling) or by incrementally selecting and spawning simulation traces with higher likelihood of reaching the rare event (e.g., importance splitting) [24]. Instead, RAGTIMER produces numerous (shortest) traces *proven* to terminate in a rare-event state, essentially performing a partial state space exploration of the model, and then computes the cumulative probability of each trace.

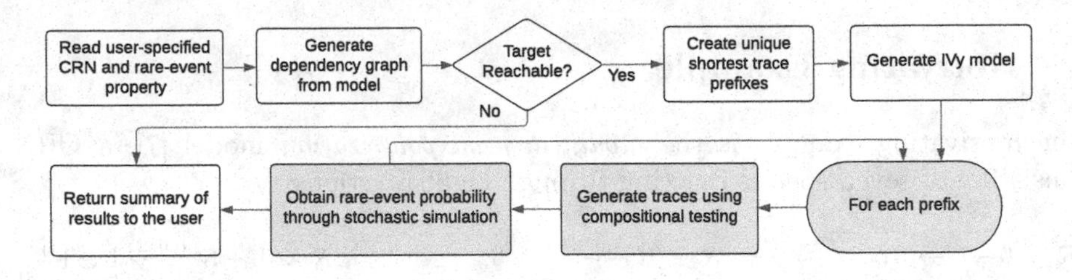

Fig. 1. RAGTIMER Flowchart

6 Layered and Service-Oriented CRN Model Generation

Conventionally, a CRN is modeled as a set of concurrently executing guarded commands, each presenting a constituent chemical reaction, such as those produced by the SBML-to-PRISM converter [1,21]. The modeling approach in this work presents a fresh perspective by considering a CRN as a layered set of service objects that maintain all of its constituent chemical reactions:

1. Layer 0 includes the following objects: `enabled_checker` to evaluate a reaction's readiness to occur, `selector` to select an enabled transition to execute, `updater` to update species as the result of a reaction, `inspector` to monitor reaction behavior, and `goal` to check reachability of the desired target.
2. Layer 1 includes a top-level object `protocol` to manage the execution of all constituent reactions of the CRN by calling services at the lower layer.

In a CRN model $\mathcal{M} = \langle \mathcal{X}, \mathcal{R}, s_0 \rangle$, every reaction $\mathcal{R}_i \in \mathcal{R}$ is modeled as an action `update_Ri` in `protocol`. Actions in `enabled_checker` check whether a given reaction has sufficient reactant(s) to occur at each state. Actions in the `selector` object can be configured to determine the frequency of executing enabled reactions. The `updater` object has actions to increment or decrement a given species according to the state change vector for $\mathcal{R}_i$. The action in the `goal` object monitors whether an execution sequence of reactions has reached a state where the goal is achieved. Note that the layered and service-oriented IVy model presented in this section omits reaction rates and hence is probability-abstract due to IVy's lack of support in floating point operations needed for computing probabilities. Acquisition of rare-event probability is described in Sect. 9.

The layered modeling approach naturally facilitates modularity and rare-event trace generation using compositional testing, which is a feature provided by the IVy verification tool [27,28]. Consider `protocol` as process p_1 as shown in Rule (2). A mirror process for `protocol` and all layer-zero objects together form the environment process p_2. The *mirror* process is one in which no actions

```
object protocol = {
 before update_R3 {
  assert enabled_checker.
   is_enabled_R3(r, l)}
 action update_R3 = {
  if selector.execute_R3 {
   call inspector.
    check_guard_R3(r, l, rl);
   r := updater.decr(r);
   l := updater.decr(l);
   rl := updater.incr(rl);
   l := updater.incr(l) }}}
```

```
object enabled_checker = {
 action is_enabled_R3(r_1:
   updater.num, l_1:updater.
   num) returns(y:bool) = {
  if r_1 >= 1 & l_1 >= 1 { y
   := true }
  else { y := false }}}
object inspector = {
 before check_guard_R3 {
  assert r_1 >= 1 & l_1 >= 1}
 action check_guard_R3(r_1:
   updater.num, l_1:updater.
   num, rl_1:updater.num)}
```

Fig. 2. IVy model snippet showing `protocol`'s `update_R3` action calling other actions in lower-layer objects to execute reaction $\mathcal{R}_3$.

are defined and its only purpose is to nondeterministically call the *exported* actions defined in p_1. An exported `idling` action is also defined in p_1, which becomes the *only* available action after the goal is achieved. It models that $\mathcal{M}$ becomes idle after a target state has been reached for the first time. This enables us to effectively curtail the model execution trace after it reaches the goal. All layer-zero objects also constituting p_2 each define actions that can only be called by p_1, effectively contributing to output actions for p_1. Rare-event trace generation is achieved by compositionally testing the triple $\langle\alpha\rangle p_1\langle\alpha\rangle$ in isolation. Isolation of p_1 (i.e., `protocol`) is with respect to its environment p_2. For correctness, checking of p_1's outputs is predicated only on p_2's assumptions. Both assumption and guarantee formulas are declared as assertions in the IVy model. When p_1 is checked in isolation, assertions taking place before p_1 action calls become assumptions, and those following p_1 action calls become guarantees. IVy's compositional testing tool then generates randomized inputs satisfying the assumptions α, while checking that the guarantees α are not violated.

We use reaction $\mathcal{R}_3$ at s_0 in the motivating example as an illustration. As Fig. 2 shows, before `update_R3` can happen, $\mathcal{M}$ first checks its precondition (expressed as an assertion) by calling action `is_enabled_R3` defined in `enabled_checker`. It checks whether $\mathcal{R}_3$ has sufficient reactants to occur at the current state. If so, it is enabled and `update_R3` calls `selector.execute_R3`, which determines whether to execute $\mathcal{R}_3$. Only when it is selected can the call be made to action `inspector.check_guard_R3`, whose `before` monitor checks the sufficient precondition again for $\mathcal{R}_3$. The assertion in `is_enabled_R3` of `enabled_checker` is converted to an assumption, but the one in `check_guard_R3` of `inspector` becomes a guarantee for action `update_R3`. Note that these two assertions are functionally equivalent. For example, they both check that the following specification α, `r_1 >= 1 & l_1 >= 1`, holds during compositional testing. When an assumption for this action fails to hold, model execution skips this action. However, violation of a guarantee will halt model execution and report a failure. This guarantee is checked by `inspector.check_guard_R3`, rather than the earlier call to action `enabled_checker.is_enabled_R3`. This is because the purpose of `is_enabled_R3` is to determine whether $\mathcal{R}_3$ is enabled to occur during model execution, but it may not be selected even if it is enabled during testing. Therefore, having a guarantee in this action similar to that of `inspector.check_guard_R3` actually leads to incorrect behavioral modeling, since the execution should not stop when $\mathcal{R}_3$ is merely disabled. However, it is necessary to guarantee sufficient reactants in `inspector.check_guard_R3`, because in order to reach this point, $\mathcal{R}_3$ must have already been selected to occur. Failure of this guarantee stops the execution, as it reveals a flaw in the model. If $\mathcal{R}_3$ successfully executes, `update_R3` updates its reactants and products by calling actions in `updater`. Lastly, the `protocol`'s environment process arbitrarily chooses another exported action to execute and this procedure repeats until the goal is achieved, after which, only the `idling` action is allowed to execute.

A *run* for $p_1 \| p_2$ is a finite sequence of action calls made by both p_1 and p_2. A *valid run* r is a finite sequence of action calls made by both p_1 and p_2 ending with $\texttt{idling}_1$ or $\texttt{goal}_2$. Recall that the subscript indicates the process in which an action is defined. Define reaction $\mathcal{R}_i$ as a finite sequence of actions with the prefix $\texttt{is\_enabled\_Ri}_2$, $\texttt{update\_Ri}_1$, $\texttt{execute\_Ri}_2$, $\texttt{check\_guard\_Ri}_2$, and followed by a continuous sequence of either $\texttt{incr}_2$ or $\texttt{decr}_2$. A (valid) *trace* σ is a finite sequence of reactions obtained by removing actions in r that do not belong to a reaction. An example of a valid run is: $\texttt{is\_enabled\_R5}_2$, $\texttt{is\_enabled\_R3}_2$, $\texttt{update\_R3}_1$, $\texttt{execute\_R3}_2$, $\texttt{check\_guard\_R3}_2$, $\texttt{decr}_2$, $\texttt{decr}_2$, $\texttt{incr}_2$, $\texttt{incr}_2$, ..., $\texttt{goal}_2$. In this run, p_2 tries to execute `update_R5`, but fails because $\mathcal{R}_5$ is disabled. It then succeeds in executing $\mathcal{R}_3$, making it the first reaction in the extracted trace. Note that our model preserves the atomicity of a single reaction execution. Utilizing the atomic `action` construct in the IVy language, action `update_Ri` requires that all actions within itself fully execute before another reaction can occur. A comprehensive description of IVy's language syntax and semantics, as well as the IVy verification tool can be found in [26,27,31].

7 Shortest Trace Generation

This section introduces the dependency graph for a CRN and describes how it either proves unreachability of a rare event or guides the shortest desirable trace generation by automated assertion creation and insertion in the IVy model, which is used to rapidly enumerate many such traces through compositional testing.

Dependency Graph for Shortest Trace Generation. Denote the edge $\rightsquigarrow$ as the *dependency relation* between two reactions, which is a binary relation defined in Definition 3. $\mathcal{R}_i \rightsquigarrow \mathcal{R}_j$ indicates that $\mathcal{R}_i$ *depends on* $\mathcal{R}_j$. Denote $\rightsquigarrow^+$ as the transitive closure of the dependency relation and $\mathcal{R}_i \rightsquigarrow^+ \mathcal{R}_j$ means that there is a path of dependency edges from $\mathcal{R}_i$ to $\mathcal{R}_j$: $\mathcal{R}_i \rightsquigarrow \cdots \rightsquigarrow \mathcal{R}_k \rightsquigarrow \cdots \rightsquigarrow \mathcal{R}_j$. Dependency relationships are established based on the minimum number of times a reaction must execute for the model to reach a target state. For consistency, create an abstract reaction $\mathcal{R}_\Psi$ representing the target specification Ψ such that $\Delta(\mathcal{R}_\Psi) = C_\Psi - s_0(\mathcal{X}_\Psi)$ as described in Definition 1. Informally, $\Delta(\mathcal{R}_\Psi)$ represents the difference between the target and initial values for the abstract species $\mathcal{X}_\Psi$, and $\mathcal{R}_\Psi$ is required to execute at least $\Delta(\mathcal{R}_\Psi)$ times to achieve the target value. This provides a starting point for dependency graph construction.

Definition 1 (Minimum Required Reaction Executions). *The minimum required reaction executions $\Delta : \mathcal{R} \to \mathbb{Z}$ maps a reaction $\mathcal{R}_i$ in the set of reactions $\mathcal{R}$ of a CRN $\mathcal{M}$ to the minimum number of times $\mathcal{R}_i$ must execute for $\mathcal{M}$ to reach a target state s_Ψ from its initial state s_0.*

Definition 2 (Enabled and Disabled Reaction Sets). *The reaction set $\mathcal{R}$ is partitioned into two subsets: set E containing only reactions $\mathcal{R}_i$ enabled to execute $\Delta(\mathcal{R}_i)$ times from s_0, and set D containing all other reactions. As such, $E \cup D = \mathcal{R}$ and $E \cap D = \emptyset$.*

Definition 3 (Dependency Relation). *A dependency relation with respect to a CRN $\mathcal{M} = \langle \mathcal{X}, \mathcal{R}, s_0 \rangle$ is an antireflexive binary relation $\rightsquigarrow \subseteq D \times \mathcal{R}$. $\mathcal{R}_i$ depends on $\mathcal{R}_j$, denoted as $\mathcal{R}_i \rightsquigarrow \mathcal{R}_j$, iff all conditions below hold:*

1. $\mathcal{R}_i \in D \wedge (\mathcal{R}_i \neq \mathcal{R}_j) \wedge \neg(\mathcal{R}_j \rightsquigarrow^+ \mathcal{R}_i)$,
2. $(\Delta(\mathcal{R}_i) > 0 \wedge \mathsf{Reactant}_i \cap \mathsf{Product}_j \neq \emptyset) \vee (\Delta(\mathcal{R}_i) < 0 \wedge \mathsf{Product}_i \cap \mathsf{Reactant}_j \neq \emptyset)$, *and*
3. $(\exists \mathcal{R}_k \ s.t.\ \mathcal{R}_j \rightsquigarrow \mathcal{R}_k) \vee (\mathcal{R}_j \in E)$.

The conditions of Definition 3 are described intuitively as follows:

1. $\mathcal{R}_i$ must be initially disabled in order to depend on another reaction and it cannot depend on itself directly or cyclically.
2. $\mathcal{R}_i$ must execute a nonzero number of times. If $\mathcal{R}_i$ requires the production (consumption) of a species, $\mathcal{R}_j$ must produce (consume) that species.
3. $\mathcal{R}_j$ must depend on another reaction, or the initial state must supply abundant species counts to enable $\mathcal{R}_j$ to execute $\Delta(\mathcal{R}_j)$ times.

An algorithm for dependency graph construction first explores Conditions 1 and 2 before removing dependency relations that fail Condition 3. Note that $\Delta(\mathcal{R}_i)$ can be negative in $\mathcal{R}_\Psi$, indicating that if a target species count is less than its initial count, it is desirable to find reactions that consume, rather than produce, the required species. Cyclic dependencies are not permitted under Definition 3. This addresses the paradox in which a reaction must execute an infinite number of times to produce enough of a species to enable itself to execute. These conditions are crucial in proving unreachability, as described later in this section.

The dependency relation between all reaction pairs guides the construction of a dependency graph, which is a directed graph with vertices representing reactions and edges representing dependency relation. The root of the graph is $\mathcal{R}_\Psi$, as no other reaction can depend on it. Algorithm 1 outlines the procedure for recursively constructing the graph following Definitions 1, 2, and 3. Figure 3a shows a case for the motivating example. Starting with $\mathcal{R}_\Psi$, establish the relation

Algorithm 1. Dependency Graph Construction

Require: $\mathcal{M} = \langle \mathcal{X}, \mathcal{R}, s_0 \rangle$, Ψ.
1: **procedure** MAIN
2: Generate $\mathcal{R}_\Psi$ and calculate $\Delta(\mathcal{R}_\Psi) = C_\Psi - s_0(\mathcal{X}_\Psi)$
3: Initialize all dependency relations to `false`
4: BUILDGRAPH($\mathcal{R}_\Psi$)
5: **if** there exists $\mathcal{R}_i$ such that $\mathcal{R}_\Psi \rightsquigarrow \mathcal{R}_i$ **then** begin trace generation.
6: **else** Ψ is unreachable; terminate.

7: **procedure** BUILDGRAPH(Reaction $\mathcal{R}_i$)
8: **if** $\mathcal{R}_i \in E$ **then return**
9: **for all** $\mathcal{R}_j$ such that $\neg(\mathcal{R}_i = \mathcal{R}_j \vee \mathcal{R}_j \rightsquigarrow^+ \mathcal{R}_i)$ **do**
10: $\delta := \Delta(\mathcal{R}_i) - \min_{\mathcal{X}_k \in \mathsf{Reactant}_i}(s_0(\mathcal{X}_k))$
11: **if** $\delta > 0$ **then** $\Delta(\mathcal{R}_j) := \Delta(\mathcal{R}_j) + \delta$; $\mathcal{R}_i \rightsquigarrow \mathcal{R}_j :=$ `true`
12: BUILDGRAPH($\mathcal{R}_j$)
13: **if** $\neg((\exists \mathcal{R}_k \text{ s.t. } \mathcal{R}_j \rightsquigarrow \mathcal{R}_k) \vee (\mathcal{R}_j \in E))$ **then** $\mathcal{R}_i \rightsquigarrow \mathcal{R}_j :=$ `false`

$\mathcal{R}_\Psi \rightsquigarrow \mathcal{R}_5$ since only $\mathcal{R}_5$ can produce G_{bg} required by $\mathcal{R}_\Psi$. The relations $\mathcal{R}_5 \rightsquigarrow \mathcal{R}_3$ and $\mathcal{R}_5 \rightsquigarrow \mathcal{R}_8$ are similarly established. Since both $\mathcal{R}_3$ and $\mathcal{R}_8$ are enabled in s_0, the algorithm terminates without removing any relations. Figure 3b, in contrast, shows an unreachable target state assuming a different s_0, where it fails to supply sufficient molecules of G, e.g., $s_0(G) = 1$. Thus, relations $\mathcal{R}_5 \rightsquigarrow \mathcal{R}_3$, $\mathcal{R}_5 \rightsquigarrow \mathcal{R}_8$, and $\mathcal{R}_5 \rightsquigarrow \mathcal{R}_7$ are established. Since $\mathcal{R}_7 \notin E$, $\mathcal{R}_7 \rightsquigarrow \mathcal{R}_6$ and $\mathcal{R}_7 \rightsquigarrow \mathcal{R}_5$ are established, creating cyclic dependency $\mathcal{R}_5 \rightsquigarrow^+ \mathcal{R}_5$. Therefore, $\mathcal{R}_7 \rightsquigarrow \mathcal{R}_5$ is disestablished. Because it is impossible to produce enough G_{bg} to execute $\mathcal{R}_7$, $\mathcal{R}_5 \rightsquigarrow \mathcal{R}_7$ is disestablished. Similarly, $\mathcal{R}_\Psi \rightsquigarrow \mathcal{R}_5$ must be disestablished, leaving only $\mathcal{R}_\Psi$ in the dependency graph, proving unreachability of Ψ.

The derivation of a dependency graph enables three crucial developments. First, it enables CRN reachability analysis as described in Theorem 1. Second, a set of desirable reactions enables the construction of traces leading to a target state. This is useful when very few traces reach a target state. Lastly, shortest traces (by reaction execution count) are obtainable as shown in Theorem 2.

Theorem 1 (Unreachable Target). *In a dependency graph constructed per Definitions 1, 2, and 3 for a CRN* $\mathcal{M} = \langle \mathcal{X}, \mathcal{R}, s_0 \rangle$, *if* $\mathcal{R}_\Psi \in D \wedge (\forall \mathcal{R}_i \in \mathcal{R}.\ \neg(\mathcal{R}_\Psi \rightsquigarrow^+ \mathcal{R}_i \wedge \mathcal{R}_i \in E)$, *any target state* s_Ψ *is unreachable from* s_0.

Proof (Theorem 1). Assume the following opposite statement holds: A dependency graph constructed with Definitions 1, 2, and 3 for a CRN satisfies $\mathcal{R}_\Psi \in D \wedge (\forall \mathcal{R}_i \in \mathcal{R}.\ \neg(\mathcal{R}_\Psi \rightsquigarrow^+ \mathcal{R}_i \wedge \mathcal{R}_i \in E)$ and some target state s_Ψ is reachable. In a closed CRN, if s_Ψ is reachable, then either (1) $\mathcal{R}_\Psi \in E$ in s_0 or (2) $\mathcal{R}_\Psi \in D$ in s_0 and there exists a reaction sequence, say $\mathcal{R}_i \ldots \mathcal{R}_l$ and $\mathcal{R}_i \in E$, that enables $\mathcal{R}_\Psi$ in a future state. Case (1) is trivially true for this theorem. For case (2), construct a dependency graph by Definition 3. Then it must hold that $\exists \mathcal{R}_i \in \mathcal{R}.\ \mathcal{R}_\Psi \rightsquigarrow^+ \mathcal{R}_i \wedge \mathcal{R}_i \in E$. Because $\mathcal{R}_i \ldots \mathcal{R}_l$ exists to enable $\mathcal{R}_\Psi$, the constructed dependency graph must establish $\mathcal{R}_\Psi \rightsquigarrow^+ \mathcal{R}_i$ (Condition 2) and the terminal reaction $\mathcal{R}_i$ for $\mathcal{R}_\Psi \rightsquigarrow^+ \mathcal{R}_i$ must be enabled in s_0

(Condition 3) when $\mathcal{R}_\Psi \rightsquigarrow^+ \mathcal{R}_i$ is acyclic (Condition 1). This contradicts the assumption. □

Grouping dependency relations into weighted branches as described in Definition 4 enables the discovery of trace lengths. Intuitively, a weighted branch $\mathcal{B}$ is a weighted path $\mathcal{R}_\Psi \rightsquigarrow^+ \mathcal{R}_i \wedge \mathcal{R}_i \in E$ in the dependency graph. The branch weight $W(\mathcal{B})$ is the minimal length of a trace including only the reactions in $\mathcal{B}$.

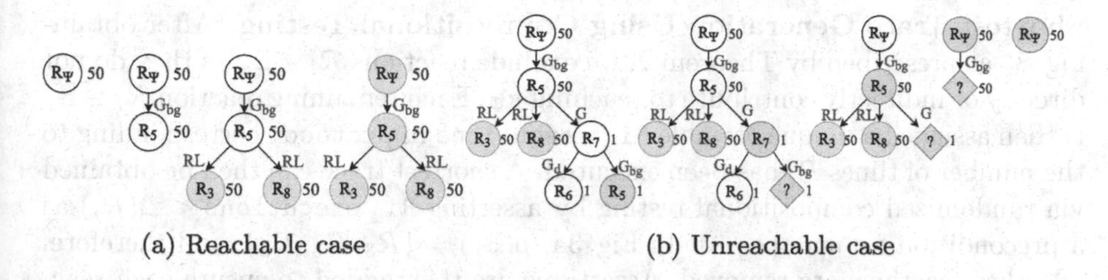

Fig. 3. Dependency Graph Examples.

Definition 4 (Weighted Branch on a Dependency Graph). *Let a branch* $\mathcal{B}_\alpha = \{\mathcal{R}_i \mid (\mathcal{R}_\Psi \rightsquigarrow^+ \mathcal{R}_i \rightsquigarrow^+ \mathcal{R}_j \wedge \mathcal{R}_j \in E) \vee (\mathcal{R}_\Psi \rightsquigarrow^+ \mathcal{R}_i \wedge \mathcal{R}_i \in E)\}$. *The weight of* $\mathcal{B}_\alpha$ *is* $W(\mathcal{B}_\alpha) = \sum_{\mathcal{R}_i \in \mathcal{B}_\alpha} \Delta(\mathcal{R}_i)$, *which is the sum of the minimal number of reaction executions when executing only the reactions in* $\mathcal{B}_\alpha$. *Label a branch with* $L(\mathcal{B}_\alpha) = \{\mathsf{Reactant}_i \cap \mathsf{Product}_j \mid \mathcal{R}_i \rightsquigarrow \mathcal{R}_j \wedge \mathcal{R}_i, \mathcal{R}_j \in \mathcal{B}_\alpha\}$.

Theorem 2 (Shortest Trace). *Construct a set of branches* $\mathbb{B}_x$ *such that each branch is required to produce a unique species and* $\forall\, \mathbb{B}_y \neq \mathbb{B}_x$, $\sum_{\mathcal{B}_a \in \mathbb{B}_x} W(\mathcal{B}_a) \leq \sum_{\mathcal{B}_b \in \mathbb{B}_y} W(\mathcal{B}_b)$. *Let* $\rho(\mathbb{B}_x) = \bigcup_{\mathcal{B}_a \in \mathbb{B}_x} \mathcal{R}_i \in \mathcal{B}_a$. *There exists some sequence of reactions in* $\rho(\mathbb{B}_x)$, *where each* $\mathcal{R}_i \in \rho(\mathbb{B}_x)$ *is executed precisely* $\Delta(\mathcal{R}_i)$ *times such that the resulting trace yields the minimal reaction execution count.*

Intuitively, a branch is a set of reactions that together produce or consume at least one desired species for $\mathcal{R}_\Psi$. In the motivating example shown in Fig. 3a, there are two branches: $\{\mathcal{R}_5, \mathcal{R}_3\}$ and $\{\mathcal{R}_5, \mathcal{R}_8\}$. Shortest traces are generated by first gathering the non-empty lowest-weighted set of branches producing unique required species. For example, $\mathcal{R}_3$ and $\mathcal{R}_8$ produce the same species needed by $\mathcal{R}_\Psi$, so only one branch is required to be included in $\mathbb{B}_x$, but both $\mathcal{R}_3$ and $\mathcal{R}_8$ can be included in $\mathbb{B}_x$ since $\Delta(\mathcal{R}_3) = \Delta(\mathcal{R}_8)$. If another branch produced a different required species, it would be included in $\mathbb{B}_x$ with $\mathcal{R}_3$ and/or $\mathcal{R}_8$.

Proof (Theorem 2). Consider a trace σ_η containing n_η reaction executions produced by the method in Theorem 2. Consider a trace σ_β containing n_β reaction executions, such that $n_\beta < n_\eta$. Because σ_β reaches a target state from the initial state, σ_β must include reactions that are required to reach a target state from the initial state. These actions are, by definition, included in a branch in the dependency graph. Because each reaction $\mathcal{R}_i$ in σ_β is, by definition, required to execute at least $\Delta(\mathcal{R}_i)$ times, the branches used to derive sequence σ_β from the dependency graph must have a lesser cumulative weight than the branches used to derive sequence σ_η. However, in constructing sequence σ_η, only the branches with the least cumulative weight are selected. Thus, either σ_β is of the same length as σ_η (causing both traces to be shortest traces), or σ_β cannot exist, proving by contradiction that a trace σ_η derived by the procedure in Theorem 2 must be a trace with the fewest total reaction executions. □

Shortest Trace Generation Using Compositional Testing. After obtaining $\mathbb{B}_x$ as prescribed by Theorem 2, we exclude reactions $\mathcal{R}_l \notin \mathbb{B}_x$ as they do not directly or indirectly contribute to reaching s_Ψ. Each remaining reaction $\mathcal{R}_i \in \mathbb{B}_x$ is then assigned a unique variable `Ri_executions` in `protocol` corresponding to the number of times $\mathcal{R}_i$ has been executed. A shortest trace can then be obtained via randomized compositional testing by asserting `Ri_executions` $< \Delta(\mathcal{R}_i)$ as a precondition for `update_Ri`. In Fig. 3a, $\rho(\mathbb{B}_x) = \{\mathcal{R}_3, \mathcal{R}_5, \mathcal{R}_8\}$, and therefore, all other reactions are removed. Assertions are then added to ensure each reaction only executes the minimum required number of times needed to reach s_Ψ. For example, assertion `R5_executions` < 50 is added to `update_R5` to ensure that $\mathcal{R}_5$ does not execute more than 50 times. During compositional testing of the triple $\langle\alpha\rangle p_1 \langle\alpha\rangle$, p_1 is the `protocol` object in a CRN model and α is `Ri_executions` $< \Delta(\mathcal{R}_i)$. This assertion serves as a part of the precondition for action `update_Ri` defined in p_1. It becomes an assumption when testing p_1 locally. Execution of every $\mathcal{R}_i \in \rho(\mathbb{B}_x)$ starts with action `update_Ri`, which is called by p_1's environment process p_2. $\mathcal{R}_i$ cannot occur if `Ri_executions` has already reached $\Delta(\mathcal{R}_i)$, indicating that the minimal number of the required $\mathcal{R}_i$'s action sequences to reach s_Ψ has already occurred, even if $\mathcal{R}_i$ is enabled. The same assertion is inserted as a part of the precondition for the `check_guard_Ri` action defined in the `inspector` object, which is part of p_2. It becomes a guarantee during the local testing of p_1 because `check_guard_Ri` is p_1's output action. Inserting the assertion `Ri_executions` $< \Delta(\mathcal{R}_i)$ in this way guarantees that only the minimum number of every $\mathcal{R}_i \in \rho(\mathbb{B}_x)$ can appear in the resulting trace σ, whose length is therefore, the shortest. Note that only once the contributing reactions (i.e., those in $\mathbb{B}_x$) have been established from the dependency graph is the IVy model actually constructed with all of the aforementioned assertions.

8 Generation of Diverse Traces

To optimize the chance of finding representative traces leading to rare-event states with relatively high probability, it is desirable to obtain a diverse assortment of traces. Using reactions from equally weighted branches and sequence prefixes can generate a diverse set of traces from compositional testing.

Equally Weighted Sets of Branches. Given $\mathbb{B}_x = \{\mathcal{B}_a, \mathcal{B}_b\}$, $\mathcal{R}_\Psi \rightsquigarrow^+ \mathcal{R}_i \rightsquigarrow^+ \mathcal{R}_x$, and $\mathcal{R}_\Psi \rightsquigarrow^+ \mathcal{R}_i \rightsquigarrow^+ \mathcal{R}_y$, if a species $\mathcal{X}_i \in \mathsf{Reactant}_i$ is produced equally effectively by a reaction $\mathcal{R}_y \in \mathcal{B}_b$ as by a reaction $\mathcal{R}_x \in \mathcal{B}_a$, in order to produce more diverse traces, the traces should include both $\mathcal{R}_x$ and $\mathcal{R}_y$, but they should be executed at a lower frequency than that of other reactions, in order to not favor the production of one species more than another during compositional testing. To accomplish this, let each reaction be assigned a *tier value*, where $tier(\mathcal{R}_\alpha) = |\{\mathcal{R}_i \in \mathcal{B} \mid (\mathcal{R}_i \rightsquigarrow^+ \mathcal{R}_\alpha\}|$ and let the *frequency* of $\mathcal{R}_\beta$ be denoted as $freq(\mathcal{R}_\beta) = |\{\mathcal{R}_\alpha \mid tier(\mathcal{R}_\alpha) = tier(\mathcal{R}_\beta) \wedge \mathcal{R}_\alpha, \mathcal{R}_\beta \in \rho(\mathbb{B}_x)\}|^{-1}$. The CRN model is then modified as follows: (1) The assertion $(\texttt{Rx\_executions} + \texttt{Ry\_executions}) < \Delta(\mathcal{R}_y)$ is inserted as a precondition for both $\mathcal{R}_x$ and $\mathcal{R}_y$; and (2) `selector` is modified to let $\mathcal{R}_x, \mathcal{R}_y$ execute with a frequency of $freq(\mathcal{R}_y)$. These modifications ensure that the generated traces are diverse and the shortest. For instance, for $\mathcal{R}_3$ and $\mathcal{R}_8$ in our motivating example, because $L(\mathcal{R}_5 \rightsquigarrow \mathcal{R}_3) = L(\mathcal{R}_5 \rightsquigarrow \mathcal{R}_8) = \text{RL}$, $\mathcal{R}_3$ and $\mathcal{R}_8$ have the same tier value, and $\Delta(\mathcal{R}_3) = \Delta(\mathcal{R}_8) = 50$, we change the precondition assertions for `update_R3` and `update_R8` to be $(\texttt{R3\_executions} + \texttt{R8\_executions}) < 50$. We also modify `selector` so that $\mathcal{R}_3$ and $\mathcal{R}_8$ execute once per every two times they are enabled.

Sequence Prefixes for Generating Diverse Traces. Due to the highly concurrent nature of CRNs, multiple unique weighted branch sets can create shortest traces according to Theorem 2. In the motivating example, the branches $\{\mathcal{R}_5, \mathcal{R}_3\}$ and $\{\mathcal{R}_5, \mathcal{R}_8\}$ each have a weight of 100. Executing a total of 100 reactions using a combination of $\mathcal{R}_5$, $\mathcal{R}_3$, and $\mathcal{R}_8$ as prescribed by the dependency graph results in the generation of a shortest trace. Dependency information is used to partition the state space into prefix-based subsets. For example, all traces generated using branches $\{\mathcal{R}_5, \mathcal{R}_3\}$ and $\{\mathcal{R}_5, \mathcal{R}_8\}$ must begin with the sequence prefix $\mathcal{R}_8, \mathcal{R}_5$; $\mathcal{R}_3, \mathcal{R}_5$; $\mathcal{R}_3, \mathcal{R}_8$; or $\mathcal{R}_8, \mathcal{R}_3$. Because each prefix is unique, there is no need to check for duplicate traces except among traces with identical prefixes. Therefore, it is possible to discard saved trace information after each prefix set is completed, saving only the sum of the probability of each trace. This saves time and allows for more trace enumeration using less memory. This state space partitioning method is achieved by using stochastic simulation to obtain the state immediately following each prefix. This state is then used in lieu of the model's initial state for compositional testing, and the number of required executions for each reaction in the prefix decrements. The prefix reaction sequence is prepended to each trace before stochastic simulation to obtain the trace's probability.

9 Tool Implementation

Ragtimer Tool Flow. We have implemented the proposed techniques in a prototype tool called RAGTIMER, so named because it is designed to be fast and carefree for a user, much like the Ragtime musical genre. As Fig. 1 shows, RAGTIMER first builds a dependency graph from the CRN model and its rare-event reachability property. The dependency graph is then used to determine reachability of the target rare-event specification and it either provides a proof of unreachability, or creates an IVy model containing only reactions in the set of branches with the lowest weight, as described in Theorem 2 and Sect. 8. To optimize the chance of a diverse set of representative traces to reach the target rare-event, RAGTIMER partitions the possible traces by shortest trace prefixes. Using information obtained from the dependency graph analysis, it automatically creates and then inserts assertions into the IVy model. RAGTIMER then invokes randomized compositional testing to produce a user-specified quantity of unique traces to the target state. Lastly, it simulates each trace to obtain its rare-event reachability probability by interfacing the PRISM probabilistic model checker [22] and sums them together to obtain the total probability of the set of traces.

Probability Acquisition. The main advantage of RAGTIMER is its effectiveness in rapidly generating a large number of the shortest desirable traces to a rare-event of interest. The lower bound for the rare-event probability $\mathsf{P}_{min}(\Diamond^{[0,T]}\ \Psi)$, where Ψ is the rare-event in the form of $\mathcal{X}_i = C$, is obtained by summing up the probabilities of all generated traces. RAGTIMER interfaces the stochastic simulation engine in PRISM to obtain probabilities for all individual traces. When branch prefixes are used for trace generation, this simulation occurs in batches, where each batch is for one equally weighted prefix from the dependency graph. An alternative approach is to construct a partial state space from the generated traces first and then interface a probabilistic model checking tool to obtain $\mathsf{P}_{min}(\Diamond^{[0,T]}\ \Psi)$. However, such construction would require an equivalent simulation effort for each trace in addition to state duplication. Our experiments showed that state space construction yielded an identical lower probability bound to trace simulation, but it used significantly more computational resources.

10 Results and Discussion

We obtained all results on a machine with an AMD Ryzen Threadripper 12-Core 3.5 GHz Processor and 132 GB of RAM, running Ubuntu Linux (v18.04.3). With one CPU and 16 GB of RAM allocated, RAGTIMER was tested on three representative CRN models. The rarity of their target specifications combined with the large state spaces can quickly overwhelm computational resources. Reaction propensities for all models presented in this section are in molecules per second.

Single Species Production-Degradation Model. This model consists of two species reacting through a production-degradation interaction [19]: $\mathcal{R}_1 : S_1 \xrightarrow{1.0} S_1 + S_2, \mathcal{R}_2 : S_2 \xrightarrow{0.025} \emptyset$. The initial molecule count for species vector (S_1, S_2) forms the initial state: $s_0 = [1, 40]$. The desired CSL property of this model is $\mathsf{P}_{=?}(\Diamond^{[0,100]}\ S_2 = 80)$. Obviously, the the shortest trace to s_{Ψ} is simply a repetition of $\mathcal{R}_1$. RAGTIMER quickly discovers this shortest trace, and alerts the user that only one shortest trace can be generated. Restriction is then loosened in the IVy model to allow extraneous reactions, i.e., $\mathcal{R}_2$, to randomly execute at a desired frequency to produce additional traces. Figure 4a shows the probability increase when these traces are generated. Generating 10,000 additional traces repeatedly took less than 2 minutes. This shows the value of controlling reaction restrictions. While obtaining a single shortest trace can be valuable, allowing many traces with extraneous reactions enables the accumulation of increased probability. In our experiments, when these probabilities begin to converge, it becomes helpful to increase the allowance for extraneous reactions to generate more diverse traces.

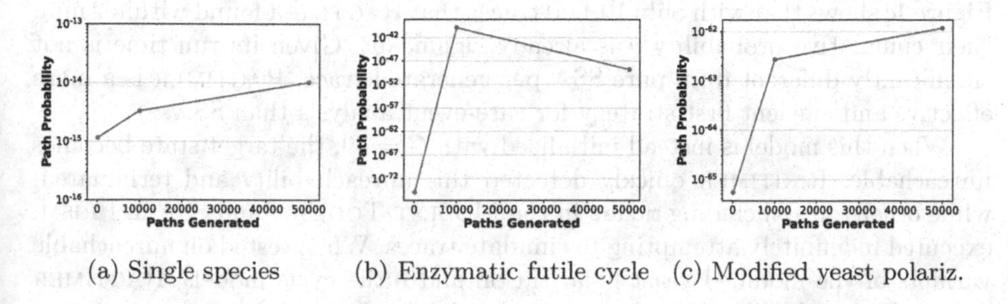

(a) Single species (b) Enzymatic futile cycle (c) Modified yeast polariz.

Fig. 4. Cumulative probability of desirable traces produced by RAGTIMER.

Enzymatic Futile Cycle Model. This example models the enzymatic futile cycle motif consisting of two single-substrate enzymatic reaction scheme, one transforming S_2 into S_5 catalyzed by S_1 and the other transforming S_5 into S_2 catalyzed by S_4 [19]:

$$\mathcal{R}_1 : S_1 + S_2 \xrightarrow{1.0} S_3, \quad \mathcal{R}_2 : S_3 \xrightarrow{1.0} S_1 + S_2, \quad \mathcal{R}_3 : S_3 \xrightarrow{0.1} S_1 + S_5,$$
$$\mathcal{R}_4 : S_4 + S_5 \xrightarrow{1.0} S_6, \quad \mathcal{R}_5 : S_6 \xrightarrow{1.0} S_4 + S_5, \quad \mathcal{R}_6 : S_6 \xrightarrow{0.1} S_4 + S_2.$$

This motif widely exists in biological systems including GTPase cycles, MAPK cascades, and glucose mobilization [37]. The initial molecule count for species vector $(S_1, S_2, S_3, S_4, S_5, S_6)$ forms the initial state: $s_0 = [1, 50, 0, 1, 50, 0]$. The rare-event property is $\mathsf{P}_{=?}(\Diamond^{[0,100]}\, S_5 = 25)$. Similar to the single-species model, this model only produces one shortest trace that alternates between $\mathcal{R}_4$ and $\mathcal{R}_6$. Figure 4b shows the probability trend with additional traces. Since these traces each reach a rare-event state, it enables a probability increase of over 30 orders of magnitude using under 2 minutes of total run time. Notably, the test with 50,000 traces yielded a much lower probability than the test with 10,000 traces, which demonstrates the importance of the user-specified frequency to execute extraneous reactions. In the test with 50,000 traces, extraneous reactions were allowed to execute more frequently, resulting in slightly longer traces with a shorter runtime per trace production. However, they lowered the overall probability despite the presence of more traces.

Modified Yeast Polarization Model. RAGTIMER is well-suited for mining rare-event traces for large-state models such as the modified yeast polarization model introduced in Sect. 4. When simulating this model using the standard *stochastic simulation algorithm* (SSA) implemented in the PRISM probabilistic model checking tool, the average trace length is generally over 4,000 reactions before reaching a target state, and the total probability for over 500,000 traces is rounded to 0 due to floating-point precision limitations. That is, SSA alone generates paths with probabilities much lower than 4.9×10^{-324}. On the contrary, RAGTIMER can find a collection of short and generally more probable traces. Figure 4c shows that with only 10,000 traces that RAGTIMER found within 2 min, their cumulative probability was already significant. Given its run time is not significantly different from pure SSA per generated trace, RAGTIMER is a more effective and efficient first strategy for rare-event analysis than SSA.

When this model is instead initialized with $G = 49$, the target state becomes unreachable. RAGTIMER quickly detected this unreachability and terminated, while other tools, including `modes` in the MODEST TOOLSET and SSA in PRISM, executed indefinitely attempting to simulate traces. When tested on unreachable variants of the modified yeast polarization and futile cycle models, RAGTIMER reported unreachability in under 1 second and did not attempt to generate traces.

Table 1. Rare-event simulation results for the three examples using `modes`.

	single species	enzymatic futile cycle	modified yeast polar
rare-event probability	3.0631×10^{-7}	1.7043×10^{-7}	1.7002×10^{-6}
runtime (seconds)	5.7	127.9	26.9
peak memory (MB)	143	141	144

Comparison to `modes` Rare-Event Simulation Engine. The `modes` tool was able to efficiently and accurately compute rare-event probabilities. Accuracy means the closeness of reported rare-event probabilities between the results from `modes`, shown in Table 1, and [19] for the single-species production-degradation and enzymatic futile cycle models, and [9] for the modified yeast polarization model. However, the compositional importance function required for rare-event simulation poses a major limitation on global variables shared among multiple components, each representing a reaction in the CRN. While manual modifications to the model's importance function can be made to get around this issue, it requires user intervention and a thorough understanding of the CRN model and the MODEST language, with the possibility of introducing errors. `modes` was also tested on the modified yeast polarization model initialized with $G = 49$ to produce an unreachable target. Despite its speed for verification of reachable models, it ran for 24 h on 23 CPU threads, performing over 1 million sweeps of the model without determining the target specification is unreachable.

Comparison to Probabilistic Model Checking Tools. After bounding all species' molecule counts to a reasonably large range of $[0, 150]$, we attempted to verify the modified yeast polarization model using the STORM probabilistic model checker [13] with the SYLVAN library [8] to allow for parallel construction of the symbolic state space. Although symbolic state space construction completed quickly, STORM failed to complete the transient analysis after running for 30 days. This is due to the overhead in converting symbolic state space to the sparse matrix representation to perform time-bounded transient analysis. Similar to STORM, PRISM was also unable to complete probabilistic model checking of this property within a reasonable time bound. However, running the state-truncation probabilistic model checker STAMINA [32] on the same model produced a probability bound of $[1.64 \times 10^{-6}, 23.01 \times 10^{-6}]$ after two days.

11 Conclusion

This paper presents a scalable and fully automated approach to rapidly enumerate a large number of traces guaranteed to reach desirable rare-event states for a given CRN model. It includes both a layered and service-oriented compositional modeling method and a dependency graph analysis technique to either prove unreachability or guide the generation of a variety of the shortest traces. Together, they automatically construct assumptions and guarantees that enable compositional testing to produce many desirable traces, which are then simulated to provide the lower probability bound for the rare-event. Efficiency in both trace generation and rare-event probability computation is demonstrated in three challenging CRN models. For future work, we will investigate effective cycle detection method to further improve rare-event trace generation.

Acknowledgment. We thank Arnd Hartmanns (U. Twente) for his help with `modes`; Chris Winstead (Utah State U.), Chris Myers (U. Colorado Boulder), and Hao Zheng (U. South Florida) for their feedback. This work was supported by the National Science Foundation under Grant No. 1856733. Any opinions, findings, and conclusions or recommendations expressed in this material are those of the authors and do not necessarily reflect the views of the funding agencies.

References

1. SBML-to-PRISM translator. http://www.prismmodelchecker.org/sbml/
2. Adelman, J.L., Grabe, M.: Simulating rare events using a weighted ensemble-based string method. J. Chem. Phys. **138**(4), 044105 (2013). https://doi.org/10.1063/1.4773892
3. Aziz, A., Sanwal, K., Singhal, V., Brayton, R.: Model-checking continuous-time Markov chains. ACM Trans. Comput. Logic **1**(1), 162–170 (2000)
4. Budde, C.E., D'Argenio, P.R., Hartmanns, A.: Automated compositional importance splitting. Sci. Comput. Program. **174**, 90–108 (2019). https://doi.org/10.1016/j.scico.2019.01.006. https://www.sciencedirect.com/science/article/pii/S0167642318301503

5. Budde, C.E., Hartmanns, A.: Replicating *Restart* with prolonged retrials: an experimental report. In: TACAS 2021. LNCS, vol. 12652, pp. 373–380. Springer, Cham (2021). https://doi.org/10.1007/978-3-030-72013-1_21
6. Chellaboina, V., Bhat, S.P., Haddad, W.M., Bernstein, D.S.: Modeling and analysis of mass-action kinetics. IEEE Control Syst. Mag. **29**(4), 60–78 (2009)
7. Daigle, B.J.J., Roh, M.K., Gillespie, D.T., Petzold, L.R.: Automated estimation of rare event probabilities in biochemical systems. J. Chem. Phys. **134**(4), 044110 (2011). https://doi.org/10.1063/1.3522769
8. Dijk, T., Pol, J.: Sylvan: Multi-core framework for decision diagrams. Int. J. Softw. Tools Technol. Transf. **19**(6), 675–696 (2017). https://doi.org/10.1007/s10009-016-0433-2
9. Donovan, R.M., Sedgewick, A.J., Faeder, J.R., Zuckerman, D.M.: Efficient stochastic simulation of chemical kinetics networks using a weighted ensemble of trajectories. J. Chem. Phys. **139**(11), 115105 (2013). https://doi.org/10.1063/1.4821167
10. Drawert, B., Lawson, M.J., Petzold, L., Khammash, M.: The diffusive finite state projection algorithm for efficient simulation of the stochastic reaction-diffusion master equation. J. Chem. Phys. **132**(7), 074101 (2010). https://doi.org/10.1063/1.3310809
11. Giannakopoulou, D., Pasareanu, C., Blundell, C.: Assume-guarantee testing for software components. Software, IET **2**, 547–562 (2009). https://doi.org/10.1049/iet-sen:20080012
12. Hartmanns, A., Hermanns, H.: The modest toolset: an integrated environment for quantitative modelling and verification. In: Ábrahám, E., Havelund, K. (eds.) TACAS 2014. LNCS, vol. 8413, pp. 593–598. Springer, Heidelberg (2014). https://doi.org/10.1007/978-3-642-54862-8_51
13. Hensel, C., Junges, S., Katoen, J.P., Quatmann, T., Volk, M.: The probabilistic model checker Storm. Int. J. Softw. Tools Technol. Transfer **24**(4), 589–610 (2022). https://doi.org/10.1007/s10009-021-00633-z
14. Jegourel, C., Legay, A., Sedwards, S.: Cross-entropy optimisation of importance sampling parameters for statistical model checking. In: Madhusudan, P., Seshia, S.A. (eds.) CAV 2012. LNCS, vol. 7358, pp. 327–342. Springer, Heidelberg (2012). https://doi.org/10.1007/978-3-642-31424-7_26
15. Jegourel, C., Legay, A., Sedwards, S.: Importance splitting for statistical model checking rare properties. In: Sharygina, N., Veith, H. (eds.) CAV 2013. LNCS, vol. 8044, pp. 576–591. Springer, Heidelberg (2013). https://doi.org/10.1007/978-3-642-39799-8_38
16. Kahn, H.: Random sampling (Monte Carlo) techniques in neutron attenuation problems-I. Nucleonics **6**(5), 27; passim (1950)
17. Kahn, H., Marshall, A.W.: Methods of reducing sample size in monte Carlo computations. J. Oper. Res. Soc. Am. **1**(5), 263–278 (1953). https://doi.org/10.1287/opre.1.5.263
18. Kahn, H., Harris, T.E.: Estimation of particle transmission by random sampling. Nat. Bureau Stand. Appl. Math. Ser. **12**, 27–30 (1951)
19. Kuwahara, H., Mura, I.: An efficient and exact stochastic simulation method to analyze rare events in biochemical systems. J. Chem. Phys. **129**(16), 165101 (2008). https://doi.org/10.1063/1.2987701
20. Kwiatkowska, M., Norman, G., Parker, D.: Stochastic model checking. In: Bernardo, M., Hillston, J. (eds.) SFM 2007. LNCS, vol. 4486, pp. 220–270. Springer, Heidelberg (2007). https://doi.org/10.1007/978-3-540-72522-0_6

21. Kwiatkowska, M., Norman, G., Parker, D.: Using probabilistic model checking in systems biology. SIGMETRICS Perform. Eval. Rev. **35**(4), 14–21 (2008). https://doi.org/10.1145/1364644.1364651
22. Kwiatkowska, M., Norman, G., Parker, D.: PRISM 4.0: verification of probabilistic real-time systems. In: Gopalakrishnan, G., Qadeer, S. (eds.) CAV 2011. LNCS, vol. 6806, pp. 585–591. Springer, Heidelberg (2011). https://doi.org/10.1007/978-3-642-22110-1_47
23. L'Ecuyer, P., LeGland, F., Lezaud, P., Tuffin, B.: Splitting techniques (2009)
24. Legay, A., Lukina, A., Traonouez, L.M., Yang, J., Smolka, S.A., Grosu, R.: Statistical model checking. In: Steffen, B., Woeginger, G. (eds.) Computing and Software Science. LNCS, vol. 10000, pp. 478–504. Springer, Cham (2019). https://doi.org/10.1007/978-3-319-91908-9_23
25. McMillan, K.: Modular specification and verification of a cache-coherent interface. In: Proceedings of the 16th Conference on Formal Methods in Computer-Aided Design, pp. 109–116. FMCAD 2016, FMCAD Inc, Austin, Texas (2016)
26. McMillan, K.L.: IVy. http://microsoft.github.io/ivy/ (2019). https://github.com/kenmcmil/ivy
27. McMillan, K.L., Padon, O.: Ivy: a multi-modal verification tool for distributed algorithms. In: Lahiri, S.K., Wang, C. (eds.) CAV 2020. LNCS, vol. 12225, pp. 190–202. Springer, Cham (2020). https://doi.org/10.1007/978-3-030-53291-8_12
28. McMillan, K.L., Zuck, L.D.: Compositional testing of internet protocols. In: 2019 IEEE Cybersecurity Development (SecDev), pp. 161–174 (2019). https://doi.org/10.1109/SecDev.2019.00031
29. Myers, C.J.: Engineering Genetic Circuits. Chapman & Hall/CRC Mathematical and Computational Biology, Chapman & Hall/CRC, 1 edn. (2009)
30. Okamoto, M.: Some inequalities relating to the partial sum of binomial probabilities. Annal. Instit. Statist. Math. **10**(1), 29–35 (1959). https://doi.org/10.1007/BF02883985
31. Padon, O., McMillan, K.L., Panda, A., Sagiv, M., Shoham, S.: Ivy: safety verification by interactive generalization. SIGPLAN Not. **51**(6), 614–630 (2016). https://doi.org/10.1145/2980983.2908118
32. Roberts, R., Neupane, T., Buecherl, L., Myers, C.J., Zhang, Z.: STAMINA 2.0: improving scalability of infinite-state stochastic model checking. In: Finkbeiner, B., Wies, T. (eds.) VMCAI 2022. LNCS, vol. 13182, pp. 319–331. Springer, Cham (2022). https://doi.org/10.1007/978-3-030-94583-1_16
33. Roh, M., Daigle, B.J.J., Gillespie, D.T., Petzold, L.R.: State-dependent doubly weighted stochastic simulation algorithm for automatic characterization of stochastic biochemical rare events. J. Chem. Phys. **135**, 234108 (2011). American Institute of Physics (2011)
34. Roh, M., Gillespie, D.T., Petzold, L.R.: State-dependent biasing method for importance sampling in the weighted stochastic simulation algorithm. J. Chem. Phys. **133**, 174106 (2010) . American Institute of Physics (2010)
35. Roh, M.K., Daigle, B.J.: SParSE++: improved event-based stochastic parameter search. BMC Syst. Biol. **10**(1), 109 (2016). https://doi.org/10.1186/s12918-016-0367-z
36. Rosenbluth, M.N., Rosenbluth, A.W.: Monte Carlo calculation of the average extension of molecular chains. J. Chem. Phys. **23**(2), 356–359 (1955). https://doi.org/10.1063/1.1741967

37. Samoilov, M., Plyasunov, S., Arkin, A.P.: Stochastic amplification and signaling in enzymatic futile cycles through noise-induced bistability with oscillations. Proceed. Nat. Acad. Sci. **102**(7), 2310–2315 (2005). https://doi.org/10.1073/pnas.0406841102. https://www.pnas.org/doi/abs/10.1073/pnas.0406841102
38. Soloveichik, D., Seelig, G., Winfree, E.: Dna as a universal substrate for chemical kinetics. Proceed. Nat. Acad. Sci. **107**(12), 5393–5398 (2010). https://doi.org/10.1073/pnas.0909380107. https://www.pnas.org/doi/abs/10.1073/pnas.0909380107
39. Villén-Altamirano, J.: Restart vs splitting: a comparative study. Perform. Eval. **121–122**, 38–47 (2018). https://doi.org/10.1016/j.peva.2018.02.002. https://www.sciencedirect.com/science/article/pii/S0166531616300839
40. Villén-Altamirano, J.: An improved variant of the rare event simulation method restart using prolonged retrials. Oper. Res. Persp. **6**, 1–9 (2019). https://doi.org/10.1016/j.orp.2019.100108. http://hdl.handle.net/10419/246387
41. Villén-Altamirano, M., Villén-Altamirano, J.: The rare event simulation method restart: efficiency analysis and guidelines for its application. In: Kouvatsos, D.D. (ed.) Network Performance Engineering. LNCS, vol. 5233, pp. 509–547. Springer, Heidelberg (2011). https://doi.org/10.1007/978-3-642-02742-0_22
42. Villen-Altamirano, M., Villen-Altamirano, J., et al.: RESTART: a method for accelerating rare event simulations. Queueing, Performance and Control in ATM (ITC-13), pp. 71–76 (1991)
43. Wald, A.: Sequential tests of statistical hypotheses. Annal. Math. Statist. **16**(2), 117–186 (1945). http://www.jstor.org/stable/2235829
44. Zhang, B.W., Jasnow, D., Zuckerman, D.M.: Efficient and verified simulation of a path ensemble for conformational change in a united-residue model of calmodulin. Proceed. Nat. Acad. Sci. **104**(46), 18043–18048 (2007). https://doi.org/10.1073/pnas.0706349104. https://www.pnas.org/doi/abs/10.1073/pnas.0706349104
45. Zuckerman, D.M., Chong, L.T.: Weighted ensemble simulation: Review of methodology, applications, and software. Annu. Rev. Biophys. **46**, 43–57 (2017). https://doi.org/10.1146/annurev-biophys-070816-033834

Accelerating Black Box Testing with Light-Weight Learning

Roi Fogler, Itay Cohen(✉), and Doron Peled

Bar Ilan University, 52900 Ramat Gan, Israel

itay.cohen5@live.biu.ac.il

Abstract. Black box testing can employ randomness for generating test sequences. Often, even a large number of test sequences may sample a minuscule portion of the overall behaviors, thus missing failures of the system under test. The challenge is to reconcile the tradeoff between good coverage and high complexity. Combining black box testing with learning (a sequence of increasingly more accurate) models for the tested system was suggested for improving the coverage of black box testing. The learned models can be used to perform more comprehensive exploration, e.g., using model checking. We present a light-weight approach that employs machine learning ideas in order to improve the coverage and accelerate the testing process. Rather than focus on constructing a complete model for the tested system, we construct a *kernel*, whose nodes are consistent with *prefixes* of test sequences that were examined so far; as part of the testing process, we keep refining and expanding the kernel. We detect whether the kernel itself contains faulty executions. Otherwise, we *exploit* the kernel to generate further test sequences that use only a *reduced* set of representative prefixes.

1 Introduction

Black box testing is far less exhaustive than model checking. Thus, it is much more likely to miss an error during testing. On the other hand, testing is often more affordable and can be used in cases where model checking is not applicable. One approach to inject learning into the black box testing process [4,11–13,16, 17] is to incrementally learn increasingly more accurate models for the tested system while performing testing. However, in many cases, the system under test has a huge state space and trying to learn first a *complete* model of the system would be impractical. Model checking can also be used on the learned models in order to cover more executions [13,16].

We incorporate into the testing process ideas from automata learning techniques, in particular the RPNI algorithm [15]. We refrain from trying to construct a complete and exact model for the system under test (SUT), due to the high complexity involved in the learning process. Instead, an approximating prefix-closed *kernel* is learned, which is consistent with the tested prefixes of the SUT.

The research was partially funded by Israeli Science Foundation grant 1464/18: "Efficient Runtime Verification for Systems with Lots of Data and its Applications".

G. Caltais and C. Schilling (Eds.): SPIN 2023, LNCS 13872, pp. 103–120, 2023.
https://doi.org/10.1007/978-3-031-32157-3_6

The kernel is iteratively extended and refined during the testing process and has the potential of growing into a complete model of the black box.

We first present an algorithm that assumes that a faulty execution enters some failure state in the SUT, which is detected during testing. We *exploit* the relation between inputs and outputs on the inspected test sequences in order to learn a kernel during the testing process. The kernel is used to reduce the length and number of prefixes that are used for generating new test cases: only the shortest paths reaching the boundary states of the kernel are extended, using randomness, to form full test sequences. This construction is in particular effective in the presence of cycles confined to the prefix, which are eliminated in forming new test sequences. The learned kernel may not be a faithful representation for prefix executions of the system due to the existence of distinguishing sequences that were not yet observed when constructing it. Thus, a controlled amount of *exploration* of alternative executions to those forming the kernel are used to incrementally refine and extend it. While a failure is not found, the testing process alternates between exploring new test cases, and constructing a refined (and larger) kernel.

Next, we present an algorithm that uses as a specification a *property automaton* for *safety properties*. Again, we alternate between exploring new test cases and constructing increasingly refined (and larger) kernels. When a new kernel is constructed, it is first model-checked against the property automaton to find whether it already contains an execution that violates the specification. Otherwise, our algorithm generates new test sequences; it again extends only a single (shortest) prefix for each boundary kernel state into full test sequences. Such an extended prefix represents all other prefixes that reach the same boundary kernel state with the same extension beyond the boundary. To facilitate using representative prefixes and find the actual failure sequence, we apply an algorithm that marks the kernel states, and states on newly generated test sequences, with the corresponding property automaton states. Then, when the property automaton indicates that a failure was detected on a test sequence, we use this marking to recover (and check whether this is a false negative) the *actual* sequence that causes the failure using a backward-forward search.

Related approaches were used in BBC (Black Box Checking) [16] and in LBtest [12,13], both using Angluin's L^* learning algorithm. Subsequent improvements of Angluin's algorithm were described in the literature [2,3,7,9,18,19,21]. In BBC and LBtest, Angluin's algorithm was used to incrementally learn a model for the system under test and perform model checking on the learned model. BBC requires the use of *equivalence queries*, which exhaustively check compatibility between the learned model and the actual black box. The complexity of these queries is exponential in the size of the black box, which is prohibitively high to be practical. LBtest (which uses additional testing strategies) relaxes the need to use equivalence queries and replaces it with approximated testing. Approximated testing was also used to replace equivalence queries in [22]. An architecture that applies a library of learning techniques in the context of testing is described in [17].

2 Preliminaries

2.1 The Model

A deterministic finite automata with output is a six-tuple $M = (S, \Sigma, \delta, \Delta, \lambda, s_0)$, where,

- S is a finite non-empty set of *states*,
- Σ is a finite non-empty *input alphabet*,
- $\delta : S \times \Sigma \rightarrow S$ is the *transition function*; we also allow the case where the successors for some states under some inputs are *undefined*, denoted by $\perp$: the inputs with *defined* successors for a state s are said to be *enabled* from s,
- Δ is the *output alphabet*, where $\Delta \cap \Sigma = \emptyset$; an automaton can distinguish a subset $F \subseteq S$ of *accepting* states, these can be, e.g., *failure states*,
- $\lambda : S \rightarrow \Delta$ is the *output function* and
- $s_0 \in S$ is the *initial state*.

An *execution* of an automaton, consists of a (finite or infinite) sequence of states, starting from the initial state s_0 and progressing according to the transition relation δ from one state to the other based on the occurrence of inputs.

A *prefix tree* or a *trie* is a search tree that is used to store associative data elements, where the keys used to find the data elements are sequences that appear on the path to the element (usually strings). We define a (prefix) tree as a six-tuple $T = (G, \Sigma, \Gamma, \Delta, w, r)$, where,

- G is the finite set of *states*,
- Σ is the finite *input alphabet*,
- Γ: $G \times \Sigma \rightarrow G$ is the *transition function*, again allowing some inputs not to be allowed out of some of the states,
- Δ is the finite *label alphabet*, where $\Delta \cap \Sigma = \emptyset$,
- $w : G \rightarrow \Delta$ is the *labeling function*, and
- $r \in G$ is the *root* state.

In our case, each path of the prefix tree corresponds to (prefix of) an execution path of a given black box automaton M that is tested, and the value of a node corresponds to the output at the end of this path; then, both Σ and Δ are the same for the automaton and the tree. Executions of M with a common prefix share that prefix when represented in a tree T.

2.2 The RPNI Algorithm

The Regular Positive and Negative Inference algorithm, or RPNI [15] is an algorithm that learns an approximation for a deterministic finite automata based on accepted and rejected sampled executions.

The RPNI algorithm consists of two phases. In the first phase, we collect a set of input sequences from Σ^* with their corresponding outputs. The algorithm builds a prefix tree [5] using paths starting from the initial state. Each tree node represents the output after the automaton executes the sequence of inputs according to the path leading to that node. In the original RPNI algorithm, the outputs are accept (1) or reject (0). It is easy to modify the algorithm so that it supports different outputs as well, which is needed in our algorithms.

In the second phase, the prefix tree is folded into an automaton by iteratively combining states that have *consistent* futures. The consistency between states g_1 and g_2 means that they cannot be distinguished by any sequence of inputs σ that exits both from g_1 and from g_2 in the prefix tree, but produces different outputs. Such sequences are called *distinguishing sequences*. In this phase, the nodes are partitioned into *red*, *blue* and *white*. Red nodes are those that were already processed. We initialize the root node to be red. Blue nodes are the immediate successors of the red nodes, and are the candidates for being combined with consistent red nodes. All other nodes are white. While not all the nodes are red, a blue node g_1 is selected and the algorithm attempts to combine it with some red node g_2. For such a combination to succeed, the algorithm checks consistency between these nodes based on distinguishing sequences. If there are no inconsistencies between g_1 and g_2, all the corresponding paths from g_1 and g_2 are merged, including the merging of g_1 and g_2 themselves. Now, either if g_1 was combined with some red g_2 node or could not be combined with any red node, g_1 becomes red and its immediate successors that are not already red become blue. The time complexity of the algorithm is cubic in the size of the prefix tree.

Note the distinction between a "combine" and a "merge" operations. The former means that the algorithm examines a red and a blue states and makes a single node out of them. The latter means that through the combination of a red and a blue state, these nodes *or some successors* of them are unified.

3 The Algorithm

3.1 The Setting

We assume that the tested system is a deterministic black box automaton with a finite (possibly unknown) number of states. Our testing procedure also works in case of an infinite state space, but in this case we cannot cover all the possibilities. The tester only knows the set of inputs Σ and outputs Δ but not the state or transition relation. The tester can *reset* the black box to its unique initial state, and then apply a sequence of inputs in order to force it to move between its states; the tester can also observe the outputs of the black box. Since not all the inputs may be enabled from all states of the black box, we assume that we obtain an indication in the form of a special output $\perp \notin \Delta$ when the tester does not succeed in progressing from the current state of the black box by trying a particular input a. We can denote this in the prefix tree as the a successor g' of g, where g' is marked with the special output $\perp$. Note that the existence of such disabled actions gives additional information that can be used by the learning algorithm.

We first present an algorithm where we assume that some states of the black box can be faulty. We do not know in advance which states are faulty, and will be able to discover them only when they are reached by following a test sequence after performing a reset.

Later, in Sect. 3.4, we change the assumption about faulty states in the black box and assume that we are given a specification in a form of a *deterministic* automaton $\mathcal{A}$. This automaton describes the bad (failed) executions as those that start at its initial state and reach a failure state $f \in F$. There is no need for F to have more than a single failure state f, and this state can be a sink. This kind of specification automaton can represent safety properties [1,8], and can be translated from past-time linear temporal logic [10]. Since an execution of $\mathcal{A}$ that reaches f is a *bad* execution, we use the SPIN [6] terminology and call $\mathcal{A}$ a *never claim*. Note that in SPIN, a never claim can be more general and also have *Büchi* accepting states, which need to occur infinitely often in order to identify an execution as faulty. Also, a never claim in SPIN can be nondeterministic.

3.2 A Modified RPNI Variant: Kernel Construction

We present a modified RPNI algorithm, which we call MRPNI.

Define the *kernel* of a prefix tree as the largest possible nonempty set of states $K \subseteq G$ such that

1. If $g \in K$ and (g', a, g) is in the prefix tree, then $g' \in K$. Thus, K is *prefix closed.*
2. For each $g \in K$, $a \in \Sigma$, there is some node $g' \in G$ such that (g, a, g') is in the prefix tree. Thus, g is *complete* with respect to its successors in the prefix tree.

During the construction of the kernel, we consider in MRPNI states to be blue, i.e. candidates for being combined with previously constructed kernel states, only if they are *complete*, i.e., all the enabled transitions from these states are explored in the prefix tree. Further, the construction guarantees that all their predecessors in the prefix tree are also complete. The reason is that merging of states in RPNI is sometimes performed based on incomplete information in the prefix tree regarding distinguishing sequences that start from these states. We use completeness as a heuristic that provides an inexpensive (but inaccurate) indication about the exploration of enough distinguishing sequences from states in the kernel; if a state is incomplete, we immediately know that some potential distinguishing sequences are unexplored from it. We stop the construction when we cannot extend the kernel further. As a result of MRPNI we obtain not only a kernel, but also paths that exit it, with some of the states outside of the kernel merged as a side effect of combining states within the kernel.

The *boundary* of a combined kernel is defined as the subset of kernel states that have a transition to at least one (incomplete) state outside the kernel. The construction of a kernel is used for optimizing the testing effort. For each state in the boundary of the kernel we exploit only a single path (the shortest one) from the initial state for generating further test cases. This path represents all the other paths in the kernel that reach this state and will be extended, using randomization, outside of the kernel for further testing.

3.3 The Basic Algorithm

We propose a procedure that assists in testing whether a faulty state is reachable from the initial state by means of a sequence of inputs. The procedure alternates between two phases. The first phase consists of calculating a prefix tree up to some point, and then obtaining a kernel of that tree using the MRPNI algorithm. The second phase uses the kernel to generate further test cases, as explained above, in order to check for reachable faulty states.

The first phase starts with building (in the first round of the two phases) or updating (in subsequent rounds) the prefix tree. We use five different conditions for switching from the first phase to the second phase, i.e., ceasing to update the prefix tree and proceeding to update the kernel.

1. A faulty state was reached when generating a new testing sequence.
2. At least half of the available outputs Δ were observed.
3. The number of new test cases explored since the previous round (unless we are in the first round) has reached a certain predefined limit.
4. An output symbol that had not been seen before was observed.
5. The number of occurrences of the least frequent observed output so far after the last MRPNI execution has increased by more than $p\%$.

Some of the conditions are heuristic and try to capture the potential for a significant progress in the search. They were driven by observations during our experiments with earlier implementations of the system. Further conditions for switching from the first phase to the second phase can be considered.

At the first round of phases, we pick up a fixed maximum length, and generate random test cases bounded by that length. The maximum length of test cases will grow from one execution of the first phase to another. We halt when one of the first three conditions 1, 2 or 3 is satisfied. If, during the construction of the prefix tree, a faulty state is reached, the process terminates, reporting the path that leads to this state.

The second phase generates new test cases based on the boundary states of the constructed kernel. We then calculate for each state g on the boundary of the kernel the shortest path σ that reaches it. We utilize the combined kernel to generate further testing cases. We select at random a boundary state g, reset the system under test and follow the shortest path σ to g. Then we extend σ with a random suffix ρ that exits the kernel, obtaining the test sequence $\sigma\rho$.

We use the reinforcement learning terminology [20] and call the construction of new test cases *exploitation* when it based on the kernel and the learning already performed. Another type of constructing test cases is called *exploration*; it is used to check whether the kernel needs to be refined. In exploration we select at random test cases, without the shortest path restrictions. These paths allow forming new distinguishing sequences: they may be used to refine the kernel in subsequent applications of MRPNI, where currently merged states in the kernel may be splitted based on newly discovered separating sequences. Note that a faulty state may be discovered, by chance, also during exploration.

We balance between exploration and exploitation by selecting the former rather than the latter with a small probability of ϵ, where the value of ϵ is refined through our experiments. We keep generating test cases as part of the second phase until either the first, the third, the fourth or the fifth condition is satisfied. If condition 1 is satisfied, we report path leading to the encountered fault and exit. If conditions 3, 4 or 5 are satisfied, we return to the first phase again, followed by the second phase and so forth. Note that condition 2 is only needed in the first round of phases.

Example. Fig. 1 shows a black box automaton with a faulty state s_7. The outputs of the state appear adjacent to the state name.

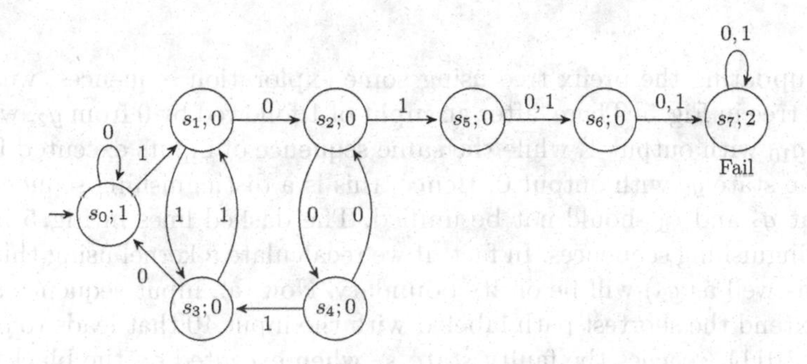

Fig. 1. A black box automaton with s_7 as its failure state.

Figure 2 illustrates a prefix tree that is generated from the black box and used as input for the MRPNI algorithm.

Figure 3 shows the prefix tree with the paired candidates to be combined encircled. The result of applying the MRPNI algorithm that combines those pairs can be seen in Fig. 4. The algorithm combines the node g_0 with g_1 (calling the unified node g_0) and the node g_2 with g_5 (calling the unified node g_2) since no sequence that distinguishes between the nodes in each pair was observed. The states of the kernel appear within a box. As a by-product of combining g_2 and g_5, g_7 is also merged with g_2 and g_5. Consequently, the resulted kernel cannot generate a sequence of inputs such as e.g. 10111 that reaches the faulty state s_7 by following the shortest path to its boundary state (which is g_2) and continuing it outside of the kernel; the shortest path is the sequence 1 (of length one), and following it outside the boundary can be done only by choosing a subsequent 1. Hence the kernel needs to be refined in subsequent rounds in order to eventually discover the fault.

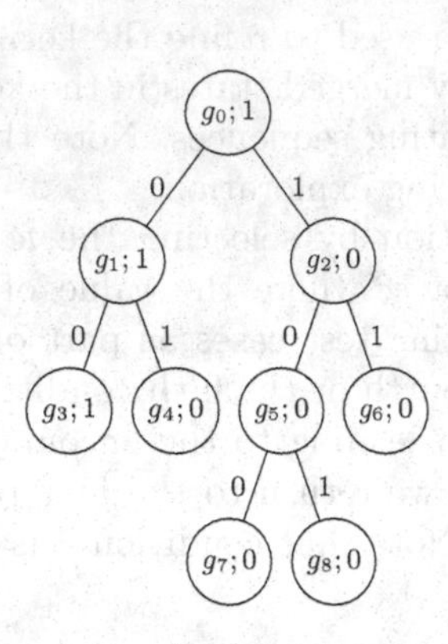

Fig. 2. A prefix tree created after generating the sequences $\{00, 01, 100, 101, 11\}$.

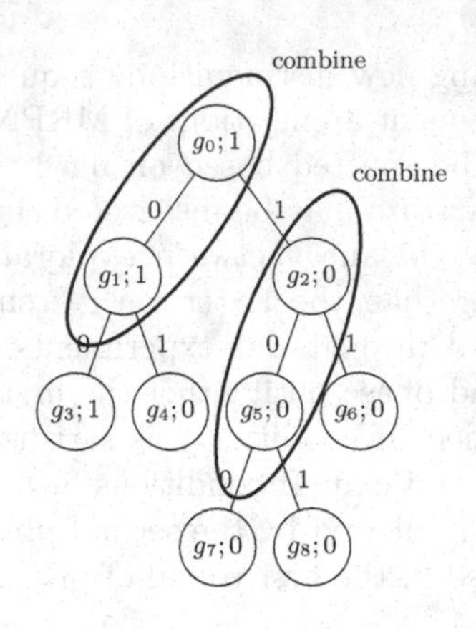

Fig. 3. candidates to be combined are g_0 with g_1 and g_2 with g_5

After updating the prefix tree using some exploration sequences, we obtain the prefix tree in Fig. 5. There, after an input of 1 followed by 0 from g_2, we reach the state g_{10} with output 1, while the same sequence of inputs executed from g_5 reaches the state g_9 with output 0. Hence, this is a distinguishing sequence that shows that g_2 and g_5 should not be unified. The dashed lines in Fig. 5 indicate these distinguishing sequences. In fact, if we recalculate a kernel using this prefix tree, g_5 (as well as g_2) will be on its boundary. Now, an input sequence such as 111 can extend the shortest path labeled with the input 10 that leads to g_5. This sequence, 10111, reaches the faulty state s_7 when executed on the black box.

3.4 Testing the Black Box Against a Specification

Annotating Kernel States in Order to Use Representative Prefixes. We describe the case where a never claim $\mathcal{A}$ is used to specify a safety property against which we want to test the black box system. As before, we want to exploit the kernel so that it is enough to follow a single path σ to a boundary state g in the kernel and then extend it using randomization outside the kernel for generating new test cases (recall that this is done under exploitation, not exploration). However, there can also be a different path σ' within the kernel leading to the same state g, where running $\mathcal{A}$ with the same inputs as σ' reaches a different state of $\mathcal{A}$ than with σ. Therefore, after constructing a new kernel, we mark each state g in it with the set of states $m(g)$ of $\mathcal{A}$ that can be reached when following *any* sequence of inputs within the kernel from the initial state r to g.

This can be done using the following procedure, based on DFS. Let Γ be the transition relation of the kernel obtained from the transition relation of the prefix tree and let ν be the transition relation of $\mathcal{A}$. Note that Γ is deterministic.

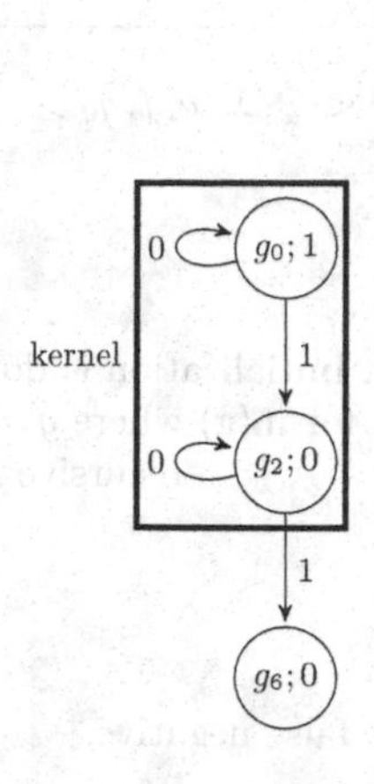

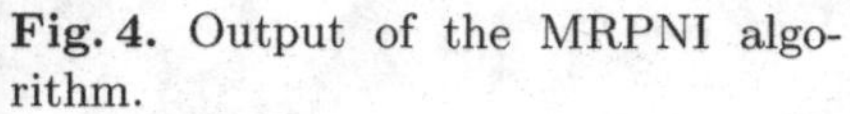
Fig. 4. Output of the MRPNI algorithm.

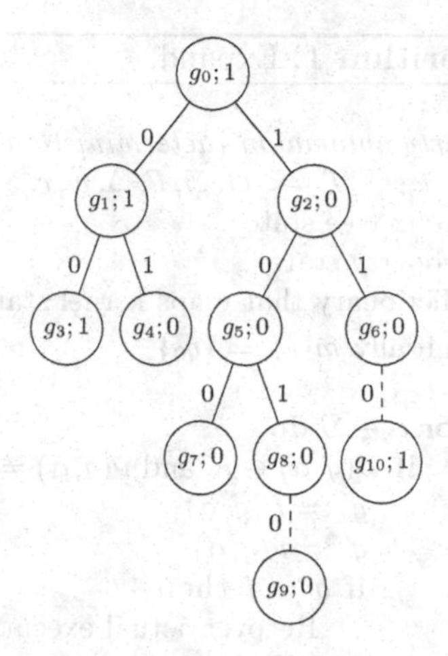

Fig. 5. Prefix tree contains distinguish sequences for nodes g_2 and g_5 as dashed lines.

We initialize $m(r)$ to $\{q_0\}$, the initial state of $\mathcal{A}$, and for any other state g in the kernel, $m(g)$ is initialized to the empty set. The search starts with the initial state r of the kernel and the initial state q_0 of $\mathcal{A}$. From a pair of states g and q of the kernel and $\mathcal{A}$, given an input $a \in \Sigma$, we move to $g' = \Gamma(g, a)$ and $q' = \nu(q, a)$ respectively, provided that such a pair of successors exists and g' is in the kernel. Then, if $q' \notin m(g')$ we set $m(g') := m(g') \cup \{q'\}$ i.e., add q' to $m(g')$ and continue recursively from g' and q'. Otherwise, we can simply backtrack. The pseudo-code for this appears below as Algorithm 1. This procedure is linear in both the size of the kernel and of $\mathcal{A}$.

If we find that for some state g in the kernel $m(g)$ contains the failure state f of $\mathcal{A}$, a failed execution that consists solely of the kernel states potentially exists. To verify this we need to find a specific path in the kernel that starts with the kernel's initial state r and reaches g such that the never claim running on the same input sequence σ' and starting with q_0 reaches f. Then we need to simulate this path on the black box to verify that this is not a false negative (obtained because of temporary combining states by the recent MRPNI, which would have been avoided given further distinguishing sequences). The case where $f \in m(g)$ is found on a state g *inside* the kernel can be considered as *model checking* the kernel. The procedure to find the aforementioned path σ' is a special case of a depth first search on the Cartesian product of the kernel and the never claim that is described later.

Note that for the simpler case, where testing for failure is based on faulty states without the use of a never claim, checking false negatives is not needed. The reason is that in this case, at the time of extending a path beyond the kernel, we follow a particular path in the black box that leads to a faulty state of the black box itself).

Algorithm 1. Expand

Input:
property automaton (deterministic finite automaton) - $P = < Q, \Sigma, \nu, f, q_0 >$
prefix tree - $T = < G, \Sigma, \Gamma, \Delta, w, r >$
g - prefix tree state
q - property state
m - dictionary that maps kernel states to property states.

```
1: Initially m(r) := {q0}        ▷ m is a global variable. Initialization is done once.
                                   Note, no initialization for m(g) where g ≠ r.
2: for α ∈ Σ do                                                  ▷ recursive function
3:     if Γ(g, α) ∈ K and ν(q, α) ≠ ⊥ then
4:         g' := Γ(g, α)
5:         q' := ν(q, α)
6:         if q' = f then
7:             Recover actual execution and check whether false negative.
8:         end if
9:         if q' ∉ m(g') then
10:            m(g') := m(g') ∪ {q'}
11:            expand(P, T, g', q', m)
12:        end if
13:    end if
14: end for
```

Consider now the case where a state g with $f \in m(g)$ was not found inside the kernel. Recall that when using the kernel to generate a new test sequence, we follow a path σ (we identify the path with the inputs since the black box is deterministic) in the kernel until reaching a boundary state g; there is a single (being in particular the shortest) path in the kernel designated for that. From g we exit the kernel along a path ρ generated at random. The state g is marked with the $\mathcal{A}$ states $m(g)$ and we need to mark the states of ρ with all the possible matching states of $\mathcal{A}$ as we have done for the states in the kernel. Let t' be the successor state of t on ρ after an input a has occurred. Then we can calculate $m(t')$ as follows (note that the first state t of ρ is on the boundary of the kernel and hence $m(t)$ is already marked).

$$m(t') = \{q' | q \in m(t) \wedge q' = \nu(q, a)\}$$

The pseudo-code of constructing ρ and marking it appears as Algorithm 2.

If $f \in m(g')$, we have reached a failure state of the never claim $\mathcal{A}$ and need to check if this corresponds to an actual execution of the black box. The algorithm for that is described below.

Recovering a Failed Execution: A Backward-Forward Search. While we are expected to benefit in efficiency and coverability by selecting to extend a single path from the boundary of the kernel, we pay with some overhead when a potential faulty execution is detected.

Algorithm 2. Subset

Input:
property automaton (deterministic finite automaton) - $\mathcal{A} =< Q, \Sigma, \nu, f, q_0 >$
prefix tree - $T =< G, \Sigma, \Gamma, \Delta, w, r >$
m - dictionary that maps prefix tree states to property states.
g - prefix tree state
α - action.

1: **if** $m(\Gamma(g, a))$ *is not defined* **then**
2: $\quad m(\Gamma(g, a)) := empty\ set$
3: **end if**
4: **for** $q \in m(g)$ **do**
5: $\quad q' := \nu(q, \alpha)$
6: $\quad$ **if** $q' = f$ **then**
7: $\quad\quad$ Recover the actual execution and check whether it is false negative.
8: $\quad$ **end if**
9: $\quad m(\Gamma(g, a)) := m(\Gamma(g, a)) \cup \{q'\}$
10: **end for**

When $f \in m(g')$, we have reached a failure state of the never claim $\mathcal{A}$, and now there are two possibilities:

1. When running the input $\sigma\rho$ on $\mathcal{A}$ from its initial state we reach the state f. Since we have simulated $\sigma\rho$ on the black box while generating it, we can immediately report this sequence as a faulty behavior.
2. We reach a state different than f. The reason is that the boundary state g at the end of the prefix σ is also the end state of another prefix, labeled with σ', of the kernel; running the input $\sigma'\rho$ on $\mathcal{A}$ would indeed reach f. The marking of $m(g)$ includes the state that $\mathcal{A}$ reaches after the input σ', although σ that may differ from σ' was used to detect the extension ρ for σ', without the need for testing σ'. In this case we need to recover σ', then check if $\sigma'\rho$ is also an admissible sequence of inputs in the black box.

Recovering σ' is performed as follows. We progress *backwards* from the end state in ρ back to g, which is its first state (and the last state of σ). We mark each state t in ρ with a set of states $l(t) \subseteq m(t)$ of $\mathcal{A}$. For the end state t of ρ we set $l(t) = \{f\}$. When progressing backwards from t to t' over an input a, meaning that $\Gamma(t', a) = t$, we set

$$l(t') = \{q' | \exists q\, \nu(q', a) = q \wedge q \in l(t)\} \cap m(t') \quad (1)$$

Thus, we trace back the set of possible states of $\mathcal{A}$ that could match the states on these paths and reach f. Note that the backward part is not needed if f is detected within the kernel.

Going backwards along the sequence ρ, we reach the state g in the kernel from which ρ started (and σ ended), with $l(g) \subseteq m(g)$. We need to find a path in the kernel labeled with σ' starting from the initial state r of the kernel, such that $\sigma'\rho$ matches a run of the never claim that starts with its initial state and ends with f. It is sufficient to search for a sequence σ' whose matching run on $\mathcal{A}$ ends with one of the states in $l(g)$. This is done using a depth first search on the Cartesian product of the kernel and the never claim $\mathcal{A}$. The search moves between pairs (g', q), where g' is a kernel state and q is a never claim state. It starts with the pair of initial states (r, q_0) and uses the successor relations Γ of the kernel and ν of the never claim to move from one such pair to another, until we reach a pair (g, q) where $q \in l(g)$.

We are not quite done yet: we still need to test the obtained path $\sigma'\rho$ against the black box to see if it is really a path or it is a false negative; the latter case can occur because of combining states by the MRPNI algorithm that would have been distinguished by further testing. In this case, we have three options:

- continue the depth first search from where it stopped, to find alternative sequence σ',
- relinquish $\sigma\rho$ and use the current kernel to search for an alternative failed execution, or
- relinquish the current kernel and perform further tests to obtain a more accurate one.

Note that in all cases, the correctness of the algorithm is preserved:

1. failures are announced only when they are not false negatives, and
2. if a failure has not been found yet, the algorithm will never block the selection of further experiments that can include failed sequences.

The interplay between these choices may affect the efficiency of the algorithm in finding errors. In particular, different choices between these cases may have the advantage over others in distinct examples.

Another difference between searching for faulty states without a never claim and the case with a never claim is in the way we select boundary states when generating exploitation test cases. While boundary states are selected randomly in the simpler case, we now select them according to a probability distribution that prioritizes states g with a relatively larger subset $m(g)$ of states of $\mathcal{A}$. The idea behind this biased selection of states is that boundary states with a larger subset of specification states are more likely to reach a failure when extended. Also, when testing a black box against a never claim, we assume that the system may not enable progressing with all the actions from each of its states. Otherwise, any sequence of $\mathcal{A}$ can be executed, which means that an execution of $\mathcal{A}$ that reaches the failure state f can always be simulated by the black box[1].

4 Experiments

We implemented our light-weight black box testing[2] and compared it with a basic random testing approach based on the average time required to reach a

[1] When testing a system without a never claim, we may assume that all the inputs are enabled from each state, or simply connect such inputs to a sink node.

[2] The code is available at https://github.com/roiDaniela/ABBT. We used the *aalpy* package [14] with some modifications to suit our specific use case.

failure. The experiments were implemented in Python and tested on the same machine with Intel® Core™ i7 CPU at 2.6 GHz and 16 GB of RAM, running on a Windows 11 operating system.

In order to compare our testing method with the basic approach, we constructed six different examples of varying characteristics and sizes. We imposed one of the states to be faulty. The state was selected according to the structure of the automaton in order to ensure a low probability of reaching the failure state. The basic random testing approach generates test sequences with a varying bounded length and determines whether the sequence results in a failure in the tested black box automaton. The test sequence actions were selected at random with equal probability.

To compare the second approach of testing the black box against a specification, we randomly generated pairs of black box and never claim automata. We produced hundreds of automata pairs of different sizes as well as input and output alphabets of varying sizes (both never claim and black box automata used the same input alphabets).

To optimize the performance of MRPNI, we added a heuristic that eliminates redundant sequences from the prefix tree. Specifically, a fraction of sequences with frequently occurring outputs was filtered out.

4.1 Examples - Basic Algorithm

The first example we experimented with is an automaton over the alphabet $\Sigma = \{0, 1\}$ that checks the divisibility of a binary number by a custom divisor. The input string for the automaton is composed of two parts. The first part is the binary representation of the divisor, which is three digits long. Hence, it can accommodate values in the range between zero to seven. The second part of the string is the binary representation of the number to be divided. The possible outputs of the automaton are $\Delta = \{n, 0, 1, 2, 3, 4, 5, 6, 7\}$. Let k be the divisor that is represented in binary form in the first part of the string. For $k \neq 0$, the output of the automaton is k if the number represented by the second part is divisible by k, and n otherwise. When $k = 0$, divisibility is not defined, and the output is arbitrarily set to be 0 for any binary number that follows the divisor in the input. We called this example *Divisibility machines.*

In the second example, *Double parentheses,* we tested an automaton that accepts strings of a balanced, nested combination of parentheses and brackets. The automaton is over the alphabet $\Sigma = \{(,), [,]\}$ and contains fourteen states. We limited the depth of the balanced parentheses in the string to four, and the depth of the balanced brackets in the string to three. The outputs in this example differ based on the characteristics of the automaton structure. The output Z corresponds to the initial state, which also represents balanced parentheses and brackets. The output C is used when a sink state is encountered. The outputs A or B is received if the last bracket/parenthesis symbol (respectively) does not exceed the depth limit.

The third example is called *Suffix one.* The automaton in this example is over the alphabet $\Sigma = \{0, 1, 2, 3\}$ and consists of twenty two states. It determines whether the symbol 1 occurs only once in the last six characters of the tested sequence. In this example, the output of each state indicates its distance (in the number of input symbols) from the initial state. While this output does not reveal much about the language that this automaton accepts, it does give insight into the structure of the automaton.

The fourth example is a black box automaton over the alphabet $\Sigma = \{0, 1, 2, 3\}$ with sixty states. This automaton accepts a string in which the number of zeros is a multiply of five, and the number of ones, twos and threes is a multiply of three each. Again, the output of each state in this example indicates its distance from the initial state. We called this example *DivBy0123*. The fifth example is called *Strings matching*. The automaton in this example is over the alphabet $\Sigma = \{a, b, c\}$, and checks if a string meets at least one of the following conditions:

- it is equal to the string "cc".
- it is equal to the string "bc".
- it is equal to the string "bcabc".
- its prefix is "ccbabcac".
- its suffix is "ab".

The outputs in this example are intended to give some information on the automaton's structure. There are five outputs, and each one corresponds to a different state responsible for detecting a different condition.

In the sixth example, *Equal stub*, we tested an automaton over the alphabet $\Sigma = \{0, 1, 2, 3\}$. The automaton checks for strings that start and end with the same substring which can be either "111", 222", or "333". The states that correspond to strings satisfying this condition have an output of 0. For all other states, the assigned output represents the number of input symbols that need to be applied from the current state until reaching a state with an output of 0.

Each example was tested 100 times against both approaches, measuring for each run the time taken until encountering a faulty state, the number of unique sequences generated, and the total sequence length generated. We report in Table 1 the average time taken, the average number of distinct sequences, and the average total sequence length.

Table 1. Experiment results - basic algorithm

Example	*Learning based testing*			*Basic random testing*		
	Time(s)	*No. seq.*	*Seq. len*	*Time(s)*	*No. seq.*	*seq. len*
Divisibility machines	0.69	3148.5	35818.8	2.2	6645.9	79009.1
Double parentheses	0.3	1771.5	11181.7	2.1	8622.6	58716.4
Suffix one	0.04	256.7	1189.0	0.2	687.5	3398.0
DivBy0123	0.1	876.5	5389.9	0.24	1521.1	9431.5
Strings matching	0.57	2490.3	18453.6	2.43	6398.3	49998.5
Equal stub	0.4	2351.9	15171.3	1.7	5617.7	37525.1

Our experiments demonstrate that our approach performs better than basic random testing in detecting failures in a black box automaton. In *Divisibility machines*, our approach reaches the failure state three times faster on average comparing to the basic random testing. The performance of our approach in *Double parentheses* is seven times faster than basic random testing. In a similar way, our approach was faster in detecting a failure in *Suffix one*. In particular, in this example our approach was approximately five times faster than the basic random testing. In contrast to the other examples, *DivBy0123* shows a smaller difference in results between the two approaches comparing to the other examples. In this example our approach was less than three times faster than the basic random testing approach. In the fifth and sixth examples, *Strings matching* and *Equal stub*, our approach reached the failure state approximately four times faster than basic random testing.

During our experiments, we have found our technique to be more effective in the cases where the probability of reaching the failure state was extremely low (rare event). This is reflected in *Double parentheses*, where there are sink states in the black box and many other sequences end in sink states. Our method may, however, be less advantageous when the failures are more dense, i.e., the probability of reaching failure states is high. This behaviour is observed in the fourth example, *DivBy0123*, where the differences between our method and basic random testing are less significant than in the other examples. In this example, our method is less than three times faster than basic random testing, as shown in Table 1.

4.2 Examples - Testing the Black Box Against a Specification

We tested our approach against four groups of examples with different sized properties. Each group contains a few dozens of randomly generated pairs of black box automaton and a never claim automaton with the same input alphabet. We attempted to find the failure in each pair of automata, using both our approach and basic random testing. The experiment for each pair was repeated ten times. The average time required to find a failure was reported, as well as the number of required distinct test sequences and their total length.

Table 2. Testing the black box against a random specification.

$\|A\|$	$\|M\|$	*Learning based testing*			*Basic random testing*		
		Time(s)	*No. seq.*	*Seq. len.*	*Time(s)*	*No. seq.*	*Seq. len.*
200	41	2.0	12481.0	306779.3	6.01	18643.5	460162.0
400	41	27.8	135134.45	4112585.6	105.6	275388.5	8481372.3
100	242	169.3	920831.8	22262209.4	729.9	1834867.0	44261905.0
40	41	0.13	1056.6	15952.6	0.34	1147.4	17270.8

The results are shown in Table 2. The first and the second groups consists of never claim automata with two hundred and four hundred states respectively, and black box automata with forty one states. In both groups it is observed that our approach outperforms basic random testing in terms of average time and number of test sequences. Our approach is three times faster in the first group and almost four times faster in the second group.

The third group exhibits a more typical case, where the never claim automaton has less states than the black box automaton itself. The advantage of our approach can also be seen in this group, as our approach is more than four times faster than basic random testing. The fourth group contains examples where both the black box and the never claim automata consist of less than fifty states. Comparing to the basic random testing, our approach was more than two times faster. However, due to the black box low complexity, both algorithms identify the failure in less than a second on average. Hence, this specific result is less significant.

When experimenting with the first group we noticed that sometimes our approach was able to find failures within the kernel while creating the subset of the never claim states that are associated with the kernel states. This enabled us to halt the procedure earlier, test less sequences and reduce the overall testing time.

We tested our approach against an additional group of examples. This group contains a few dozens of examples, and its black box automata were generated as before. However, we restricted the random generation of the never claim automata to be in the form of a "combination lock", e.g., each generated automaton has a single accepting path. The results of this experiment are presented in Table 3.

Table 3. Testing the black box against a combination lock.

$\|A\|$	$\|M\|$	*Learning based testing*			*Basic random testing*		
		Time(s)	*No. seq.*	*Seq. len.*	*Time(s)*	*No. seq.*	*Seq. len.*
20	50	17.94	743141.96	20321421.15	20.77	1223899.3	32504014.6

We observed that in this group, our approach has a less significant advantage over the basic random testing. This is only to be expected, as a combination lock is an extreme case for random testing (after all, a combination lock is designed to defy discovering it).

5 Conclusions

We presented a method to employ machine learning principles, in particular the RPNI algorithm, into probabilistic black box testing. This involves a light-weight construction and an implementation that improves the efficiency and effectiveness of black box testing. Instead of constructing a full model for the tested system, we construct kernels, which correspond to prefixes of the executions (although some infinite executions may be confined to the kernel by construction). States of the kernel can represent multiple (often many) prefixes of test cases that later confluent. Our algorithm exploits this in order to use only representatives of the prefixes in generating new test cases. We also show how to apply this method in the presence of a safety property that is represented using an automaton in the style of a never claim.

Our experiments showed positive results in the tested examples compared to basic black box testing. In both the faulty state setting and the extended setting in which the black box is tested against a specification, our approach outperforms the basic random testing approach. In both settings, the differences between the two approaches were less significant when the failure occurs with high probability.

The constructed kernel does not increase the probability to encounter a faulty execution that extends beyond the kernel (there would be fewer tested executions, but also fewer faulty executions, reduced with the same ratio). However, avoiding cycles that are confined to the kernel may be beneficial for focusing and speeding up the testing process, since shorter representatives without state repetition within the kernel can replace them and be extended to faulty executions. Another advantage of constructing a kernel becomes apparent when our black box is tested against a specification automaton $\mathcal{A}$. In this case, a potential failure can already be detected (but needs to be tested against the black box to rule out a false negative) by marking the states of the kernel with sets of states from $\mathcal{A}$ that can correspond to an execution that leads to that kernel state.

The BBC algorithm [16] and LBtest [12,13] use Angluin's L^* [2] for learning a model for the tested system. Then the model can be used for model checking. The model is iteratively refined through the testing (and model checking) process. While BBC uses expensive equivalence queries, LBtest avoids it by using non-comprehensive testing, which is more feasible. As opposed to these approaches, our algorithm learns a kernel of the black box system, corresponding to its prefixes rather than a complete model. We use the kernel for optimizing the process of performing further testing, but also, as noted, in effect perform model checking on the executions that are confined to be in the kernel. As further work, additional optimization techniques can be combined with our algorithm. It would also be interesting to compare the effectiveness of LBtest and our method using common examples involving the same black box systems and property automata.

References

1. Alpern, F.B.B.: Schneider: recognizing safety and liveness. Distrib. Comput. **2**, 117–126 (1987). https://doi.org/10.1007/BF01782772
2. Angluin, D.: Learning Regular Sets from Queries and Counterexamples. Inf. Comput. **75**, 87–106 (1987)
3. Angluin, D.: A note on the number of queries needed to identify regular languages. Inf. Control **51**, 76–87 (1981)

4. Groce, A., Peled, D., Yannakakis, M.: Adaptive model checking. Logic J. IGPL **14**, 729–744 (2006)
5. Higuera, C.: Grammatical inference: learning automata and grammars. Cambridge University Press (2010)
6. Holzmann, G.J.: The spin model checker: primer and reference manual. Addison-Wesley Professional (2014)
7. Isberner, M., Howar, F., Steffen, B.: The TTT algorithm: a redundancy-free approach to active automata learning. In: Bonakdarpour, B., Smolka, S.A. (eds.) RV 2014. LNCS, vol. 8734, pp. 307–322. Springer, Cham (2014). https://doi.org/10.1007/978-3-319-11164-3_26
8. Lamport, L.: What good is temporal logic? In: Proceedings of the IFIP 9th World Computer Congress, Information Processing, vol. 83, pp. 657–668 (1983)
9. Leucker, M.: Learning meets verification. In: de Boer, F.S., Bonsangue, M.M., Graf, S., de Roever, W.-P. (eds.) FMCO 2006. LNCS, vol. 4709, pp. 127–151. Springer, Heidelberg (2007). https://doi.org/10.1007/978-3-540-74792-5_6
10. Z. Manna, A. Pnueli, Temporal verification of reactive systems - safety, 1 Edn. Springer (1995). https://doi.org/10.1007/978-1-4612-4222-2
11. Meinke, K., Sindhu, M.A.: Incremental learning-based testing for reactive systems. In: Gogolla, M., Wolff, B. (eds.) TAP 2011. LNCS, vol. 6706, pp. 134–151. Springer, Heidelberg (2011). https://doi.org/10.1007/978-3-642-21768-5_11
12. Meinke, K., Sindhu, M.: LBTest: a learning-based testing tool for reactive systems. In: 2013 IEEE Sixth International Conference on Software Testing, Verification and Validation, pp. 447–454 (2013)
13. Meinke, K., Niu, F., Sindhu, M.: Learning-based software testing: a tutorial. In: Hähnle, R., Knoop, J., Margaria, T., Schreiner, D., Steffen, B. (eds.) ISoLA 2011. CCIS, pp. 200–219. Springer, Heidelberg (2012). https://doi.org/10.1007/978-3-642-34781-8_16
14. Muskardin, E., Aichernig, B., Pill, I., Pferscher, A., Tappler, M.: AALpy: An active automata learning library. Innov. Syst. Softw. Eng. 18, 417–426 (2022). https://doi.org/10.1007/s11334-022-00449-3
15. Oncina, J., García, P.: Inferring regular languages in polynomial updated time, series in machine perception and artificial. Intelligence **4**, 49–61 (1992)
16. Peled, D., Vardi, M., Yannakakis, M.: Black box checking. In: Proceedings of the 14th International Symposium on Mathematical Foundations of Computer Science, vol. 1672, pp. 225–240 (1999)
17. Raffelt, H., Merten, M., Steffen, B., Margaria, T.: Dynamic testing via automata learning. Int. J. Softw. Tools Technol. Transfer **11**(4), 307–324 (2009)
18. Rivest, R., Schapire, R.: Inference of finite automata using homing sequences. Inf. Comput. **103**(2), 299–347 (1993)
19. Sindhu, M.A., Meinke, K.: IDS: an incremental learning algorithm for finite automata, CoRR, vol. abs/1206.2691, pp. 1–12 (2012)
20. Sutton, R., Barto, A.: Reinforcement learning - an introduction. MIT Press, Adaptive Computation and Machine Learning (1998)
21. Vaandrager, F., Garhewal, B., Rot, J., Wißmann, T.: A new approach for active automata learning based on apartness. CoRR, vol. abs/2107.05419 (2021)
22. Weiss, G., Goldberg, Y., Yahav, E.: Extracting automata from recurrent neural networks using queries and counterexamples. In: Proceedings of the 35th International Conference on Machine Learning (ICML 2018), vol. 80, pp. 5244–5253 PMLR (2018)

Synthesis

WikiCoder: Learning to Write Knowledge-Powered Code

Théo Matricon[1(✉)], Nathanaël Fijalkow[1,2], and Gaëtan Margueritte[1]

[1] CNRS, LaBRI, University of Bordeaux, Bordeaux, France
theomatricon@gmail.com
[2] MIMUW, University of Warsaw, Warsaw, Poland

Abstract. We tackle the problem of automatic generation of computer programs from a few pairs of input-output examples. The starting point of this work is the observation that in many applications a solution program must use external knowledge not present in the examples: we call such programs knowledge-powered since they can refer to information collected from a knowledge graph such as Wikipedia. This paper makes a first step towards knowledge-powered program synthesis. We present WikiCoder, a system building upon state of the art machine-learned program synthesizers and integrating knowledge graphs. We evaluate it to show its wide applicability over different domains and discuss its limitations. WikiCoder solves tasks that no program synthesizers were able to solve before thanks to the use of knowledge graphs, while integrating with recent developments in the field to operate at scale.

Keywords: Program Synthesis · Knowledge Graphs · Code generation

1 Introduction

Task automation is becoming increasingly important in the digital world we live in, yet writing software is still accessible only to a small share of the population. Program synthesis seeks to make coding more reliable and accessible by developing methods for automatically generating code [10]. For example, the FlashFill system [9] in Microsoft Excel makes coding more accessible by allowing nontechnical users to synthesize spreadsheet programs by giving input-output examples, and TF-coder [20] assists developers with writing TensorFlow code for manipulating tensors.

Very impressive results have been achieved in the past five years employing machine learning methods to empower program synthesis. Recent works have explored engineering neural networks for guiding program search [1,5–8,16,17,19,24,27] effectively by training the network to act as a language model over source code. The most resounding success of this line of work is OpenAI's Codex system [4] powering Github's Copilot and based on very large language models. However, because it works at a purely syntactic level, program synthesis fails in a number of applications.

Let us consider as an example the following task:

G. Caltais and C. Schilling (Eds.): SPIN 2023, LNCS 13872, pp. 123–140, 2023.
https://doi.org/10.1007/978-3-031-32157-3_7

```
f("17", "United States") = "17 USD"
f("42", "France") = "42 EUR"
```

The task is specified in the programming by example setting: the goal is to construct a function f mapping inputs to their corresponding outputs. Solving this task requires understanding that the second inputs are countries and mapping them to their currencies. This piece of information is not present in the examples and therefore no program synthesis tool can solve that task without relying on external information. In other words, a solution program must be *knowledge-powered*! An example knowledge-powered program yielding a solution to the task above is given below in a Python-like syntax:

```
def f(x,y):
    return x + " " + CurrencyOf(y)
```

It uses a function CurrencyOf obtained from an external source of knowledge.

Knowledge-powered program synthesis extends classical program synthesis by targetting knowledge-powered programs. The challenge of combining syntactical manipulations performed in program synthesis with semantical information was recently set out by [22]. They discuss a number of applications: string manipulations, code refactoring, and string profiling, and construct an algorithm based on very large language models (see related work section).

Our approach is different: the methodology we develop relies on knowledge graphs, which are very large structured databases organising knowledge using ontologies to allow for efficient and precise browsing and reasoning. There is a growing number of publicly available knowledge graphs, for instance Wikidata.org, which includes and structures Wikipedia data, and Yago [12,21], based on Wikidata and schema.org. We refer to [14] for a recent textbook and to [13] for an excellent survey on knowledge graphs and their applications, and to Fig. 1 for an illustration.

The recent successes of both program synthesis and knowledge graphs suggest that the time is ripe to combine them into the knowledge-powered program synthesis research objective.

Our contributions:

- We introduce knowledge-powered program synthesis, which extends program synthesis by allowing programs to refer to external information collected from a knowledge graph.
- We identify a number of milestones for knowledge-powered program synthesis and propose a human-generated and publicly available dataset of 46 tasks to evaluate future progress on this research objective.
- We construct an algorithm combining state of the art machine learned program synthesizers with queries to a knowledge graph, which can be deployed on any knowledge graph.
- We implement a general-purpose knowledge program synthesis tool WikiCoder and evaluate it on various domains. WikiCoder solves tasks previously unsolvable by any program synthesis tool, while still operating at scale by integrating state of the art techniques from program synthesis.

2 Knowledge-Powered Programming by Example

2.1 Objectives

Programming by example, or inductive synthesis, is the following problem: given a few examples, construct a program satisfying these examples. This particular setting in program synthesis, where the user gives a very partial specification, has been extremely useful and successful for automating tasks for end users, the prime example being FlashFill for performing string transformations in Excel [9].

When considering knowledge-powered programming by example, we do not change the problem, only the solution: instead of classical programs performing syntactic manipulations, we include knowledge-powered programs. To illustrate the difference, let us consider the following two tasks.

```
f("Paris") = "I love P"
f("Berlin") = "I love B"

g("Donald Knuth") = "DK is American"
g("Ada Lovelace") = "AL is English"
```

The first is a classical program synthesis task in the sense that it is purely syntactical, it can be solved with a two-line program concatenating "I love" with the first letter of the input. On the other hand, the second requires some knowledge about the input individuals, here their nationality: one needs a knowledge-powered program to solve this task. Since almost all program synthesis tools perform only syntactical manipulations of the examples (we refer to the related work section for an in-depth discussion), they cannot solve the second task.

Knowledge-powered programming by example goes much beyond query answering: the goal is not to answer a particular query, but to produce a program able to answer that query for any input. This is computationally and conceptually a much more difficult problem.

For concreteness we introduce some terminology about knowledge graphs, and refer to Fig. 1 for an illustration. Nodes are called entities, and edges are labelled by a relation. Entities are arranged into classes: "E. Macron" belongs to the class of people, and "France" to the class of countries. The classes and relations are constrained by ontologies, which define which relations can hold between entities. The de facto standard for querying knowledge graphs is through `SPARQL` queries, which is a very powerful and versatile query language.

2.2 Milestones

We identify three independent ways in which semantical information can be used. They correspond to different stages for solving a task:

- *preprocessing:* the first step is to extract entities from the examples;
- *relating:* the second step is to relate entities in the knowledge graph;
- *postprocessing:* the third step is to process the information found in the knowledge graph.

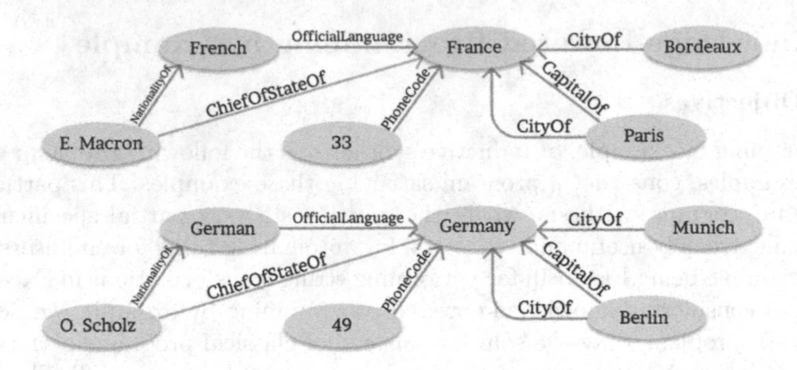

Fig. 1. Illustration of part of a knowledge graph.

How much preprocessing to extract entities? Let us consider two tasks.

```
f("Aix, Paris, Bordeaux") = "Paris"
f("Hamburg, Berlin, Munich") = "Berlin"

g("President Obama") = "Obama"
g("Prime Minister de Pfeffel Johnson") = "de Pfeffel Johnson"
```

In the function `f` the goal is to extract the second word as separated by commas: this is a purely syntactic operation. In the function `g` we need to remove the job title from the input: this requires semantical knowledge, for instance it is not enough to use neither the second nor the last word.

How complicated is the relationship between entities? We examine two more tasks.

```
f("Paris") = "France is beautiful"
f("Berlin") = "Germany is beautiful"

g("Paris") = "Phone country code: 33"
g("Berlin") = "Phone country code: 49"
```

The function `f` relates two entities: a city and a country. One can expect that the knowledge graph includes the relation `CapitalOf`, which induces a labelled edge between "Paris" and "France" as well as "Berlin" and "Germany" (as in Fig. 1). Note that it could also be the relation `CityOf`. More complex, the function `g` requires crossing information: indeed to connect a city to its country code, it is probably required to compose two relations: `CapitalOf` and `PhoneCode`. In other words, the entities are related by a path of length 2 in the knowledge graph, that we write `CapitalOf-PhoneCode`. More generally, the length of the path relating the entities is a measure of complexity of a task.

How much postprocessing on external knowledge? Let us look again at two tasks.

```
f("Paris") = "Country's first letter: F"
f("Berlin") = "Country's first letter: G"

g("President Obama") = "B@r@ck"
g("Prime Minister Johnson") = "BOr1s"
```

For the function f, the difficulty lies in finding out that the external knowledge to be search for relates "Paris" to "France" and "Berlin" to "Germany", and then to return only the first letter of the result. Similarly, for g, before applying a leet translation we need to retrieve as external knowledge the first name. In both cases the difficulty is that the intermediate knowledge is not present in the examples.

The three steps can be necessary, the most complex tasks involve at the same time subtle preprocessing to extract entities, complicated relationships between entities, and significant postprocessing on external knowledge.

2.3 Motivating Examples

We illustrate the disruptive power of knowledge-powered program synthesis in three application domains.

General Knowledge. The first example, which we use throughout the paper for illustrations, in the dataset and in the experiments, is to use the knowledge graph to obtain general facts about the world, such as geography, movies, people. Wikidata and Yago are natural knowledge graphs candidates for this setting. This domain is heavily used for query answering: combining it with program synthesis brings it to another level, since programs generalize to any input.

Grammar Exercises. The second example, inspired from [22], is about language learning: tasks are grammar exercises, where the goal is to write a grammatically correct sentence. Here the knowledge graph includes grammatical forms and their connections, such as verbs and their different conjugated forms, pronouns, adjectives, and so on. Generating programs for solving exercises opens several perspectives, including generating new exercises as well as solving them automatically.

Advanced Database Queries. In the third example knowledge-powered program synthesis becomes a powerful querying engine. This scenario has been heavily investigated for SQL queries [23,26], but only at a syntactic level. Being able to rely on the semantic properties of the data opens a number of possibilities, let us illustrate them on an example scenario. The knowledge graph is owned and built by a company, it contains immutable data about products. The database contains customer data. Crossing semantical information between the database being queried and the knowledge graph allows the user to generate complex queries to extract more information from the database including for instance complex statistics.

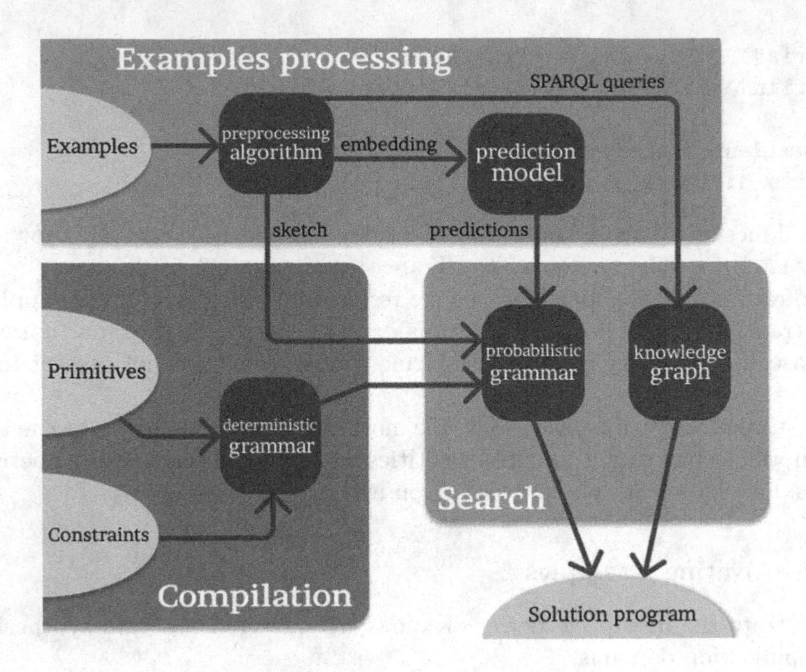

Fig. 2. WikiCoder pipeline.

3 WikiCoder

WikiCoder is a general-purpose grammar-based program synthesis tool, it was developed in Python integrating state of the art program synthesis techniques. It is publicly available on GitHub[1]. The main functionality of WikiCoder is to solve programming by example tasks: the user inputs a few examples and WikiCoder synthesizes a knowledge-powered program satisfying the examples. The programming language is specified as a domain specific language (DSL), designed to solve a common set of tasks. WikiCoder supports a number of classical DSLs, including towers building [7], list integer manipulations [1], regular expressions, and string manipulation tasks à la FlashFill [9]: our experiments report on the latter DSL. Following recent advances in machine learned program synthesizers [7,8], WikiCoder is divided into three components: compilation, examples processing, and search, as illustrated in Fig. 2 and discussed in the next three sections. This is very different from Codex's architecture, which is based on auto-regressive very large models directly generating code.

3.1 Compilation

The DSL is specified as a list of primitives together with their types and semantics. The compilation phase consists in obtaining an efficient representation of

[1] https://github.com/nathanael-fijalkow/ProgSynth/tree/WikiCoder.

the set of programs in the form of a context-free grammar. The right way to look at the grammar is as a generating model: it generates code. Thanks to the expressivity of context-free grammars, many syntactic properties can be ensured: primarily and most importantly, all generated programs are correctly typed.

The DSL we use in our experiments is tailored for string manipulation tasks à la Flashfill. For the sake of presentation we slightly simplify it. We use two primitive types: STRING and REGEXP, and one type constructor Arrow. We list the primitives below, they have the expected semantics.

```
$        : REGEXP                         # end of string
.        : REGEXP                         # all
[^_]+    : Arrow(STRING, REGEXP)          # all except X
[^_]+$   : Arrow(STRING, REGEXP)          # all except X at the end
compose : Arrow(REGEXP, Arrow(REGEXP, REGEXP))

concat   : Arrow(STRING, Arrow(STRING, STRING))
match    : Arrow(STRING, Arrow(REGEXP, STRING))

# concat_if: concat if the second argument (constant)
#            is not present in the first argument
concat_if    : Arrow(STRING, Arrow(STRING, STRING))
# split_fst: split using regexp, returns first result
split_fst  : Arrow(STRING, Arrow(REGEXP, STRING))
# split_snd: split using regexp, returns second result
split_snd : Arrow(STRING, Arrow(REGEXP, STRING))
```

In the implementation we use two more primitive types: CONSANT_IN and CONSTANT_OUT, which correspond to constants in the inputs and in the output. This is only for improving performances, it does not increase expressivity. Some primitives are duplicated to use the two new primitive types.

3.2 Examples Processing

A preprocessing algorithm produces from the examples three pieces of information: a sketch, which is a decomposition of the current task into subtasks, a set of SPARQL queries for the knowledge graph, and an embedding of the examples for the prediction model.

Preprocessing Algorithm. Figure 3 illustrates the preprocessing algorithm in action. The high level idea is that the algorithm is looking for a shared pattern across the different examples. In the task above, "code:" is shared by all examples, hence it is extracted out. A naive implementation is to look for the largest common factor between the strings, and proceed recursively on the left and on the right. The process is illustrated with Algorithm 1, which produces a list of constants from a list of strings. This procedure is applied to both the inputs and outputs independently to create sketches. Thanks to these constants

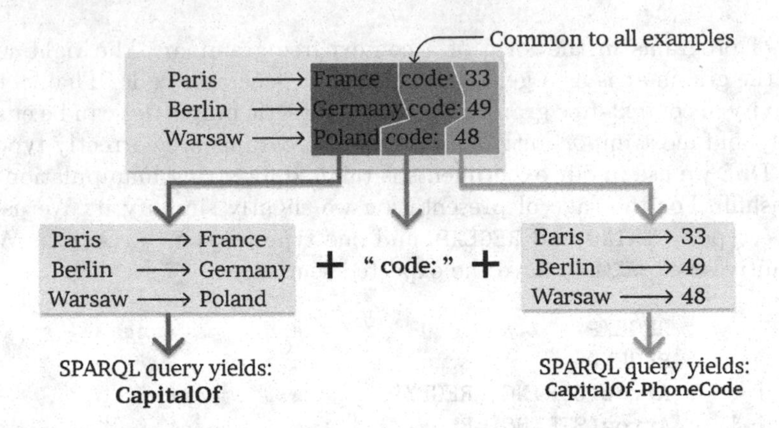

Fig. 3. Example of the preprocessing algorithm.

the task is split into subtasks as illustrated in Fig. 3 where we have split the task into two subtasks: `CapitalOf` and `CapitalOf-PhoneCode`. To solve each subtask we query the knowledge graph with the inputs and outputs. If no path is found, we run a regular – syntactical – program synthesis algorithm.

Algorithm 1. Constant extraction

procedure GETCONSTANTS($S = (S_k)_{k \in [1,n]}$: strings)
  if there is an empty string in S **then**
    return empty list
  end if
  factor ← longest common factor among all strings in S
  if len(factor) ≤ 2 **then**
    return empty list
  else
    $S_{\text{left}} \leftarrow$ prefix of factor in S
    $L_{\text{left}} \leftarrow$ GETCONSTANTS(S_{left})
    $S_{\text{right}} \leftarrow$ suffix of factor in S
    $L_{\text{right}} \leftarrow$ GETCONSTANTS(S_{right})
    return L_{left} + factor + L_{right}
  end if
end procedure

Generated SPARQL Queries. The SPARQL queries are generated from the examples after preprocessing. Once we have the constants with Algorithm 1 of the inputs and outputs, we can split the inputs and outputs in constant parts and non-constant parts, only the non-consant parts are relevant. For each non-constant part in the outputs, we generate queries from the non constant part of the inputs which should map to this non-constant part of the output. Since relations may be complex, that is "Paris" is at distance 1 from "France" but "33" is two relations away from "Paris", we generate SPARQL queries for increasing distances up to a fixed upper bound. Here is the query at distance 2, that we execute for the example in Fig. 3 with CapitalOf-PhoneCode:

```
PREFIX w: <https://en.wikipedia.org/wiki/>
SELECT ?p0 ?p1 WHERE {
    w:Paris ?p0 ?o_1_0 .
    ?o_1_0 ?p1 w:33 .
    w:Berlin ?p0 ?o_2_0 .
    ?o_2_0 ?p1 w:49 .
    w:Warsaw ?p0 ?o_3_0 .
    ?o_3_0 ?p1 w:48 .
}
```

Notice that intermediary entities make an apparition in order to accommodate for longer path lengths. The output of the above query would consist of two paths: CityOf-PhoneCode and CapitalOf-PhoneCode.

As disambiguation strategy (inspired by [25]) we choose the path with the least number of hits across all examples, called the least ambiguous path. In this example there is no preferred path since both paths lead to a single entity for each example. As an example of this disambiguation strategy, let us consider paths from "33" to "Paris": there are two paths, PhoneCodeOf-Capital and PhoneCodeOf-City. The least ambiguous path is PhoneCodeOf-Capital since PhoneCodeOf-City leads to all cities of the country.

To find the least ambiguous path, we need to count the number of hits, which is done using more SPARQL queries. Here is a sample query to get all entities at the end of the path CapitalOf-PhoneCode from the starting entity "Paris":

```
PREFIX w: <https://en.wikipedia.org/wiki/>
SELECT ?dst WHERE {
    w:Paris w:CapitalOf ?e0 .
    ?e0 w:PhoneCode ?dst .
}
```

We count the number of results for all examples to get the number of hits for a path.

Prediction Model. Efficient search is crucial in program synthesis, because the search space combinatorially explodes. We, and we believe the community more broadly, see learning as the right way of addressing this challenge. Following [7,8], we build a prediction model in the form of a neural network: it reads embedding of the examples and outputs probabilities on the derivation rules of the grammar representing all programs. This transforms the context-free grammar representing programs in a probabilistic context-free grammar, which is a stochastic process generating programs. In other words, it defines a probabilistic distribution over programs, and the prediction model is trained to maximise the probability that a solution program is generated.

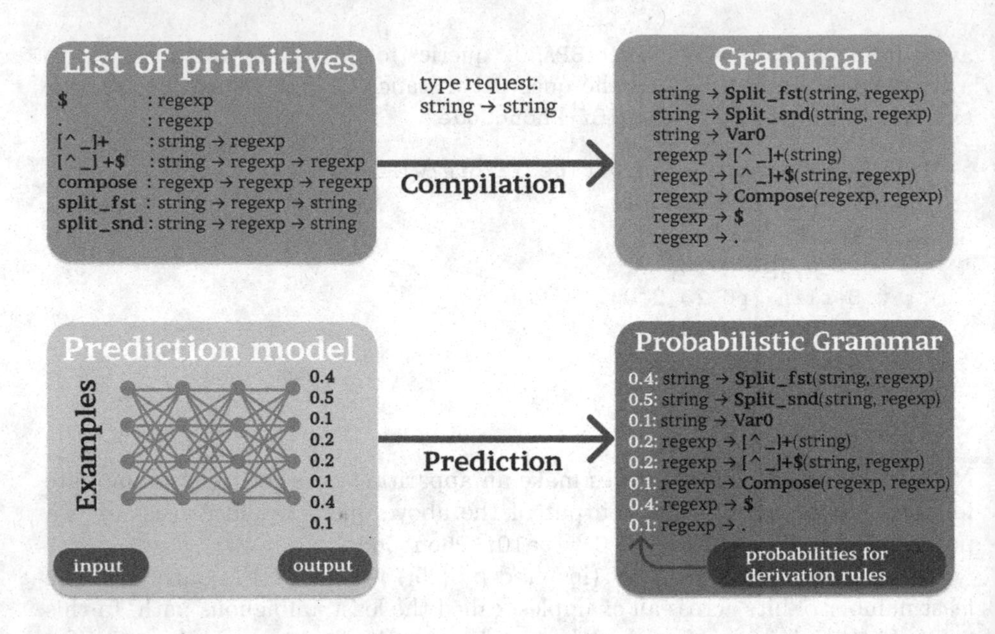

Fig. 4. Illustration of the prediction model.eps

This prediction process is illustrated in Fig. 4. This effectively leverages the prediction power of neural networks without sacrificing correctness: the prediction model biases the search towards most likely programs but does not remove part of the search tree, hence – theoretically – a solution will be found if there exists one.

3.3 Search

The prediction model assigns to each candidate program a likelihood of being a solution to the task at hand. WikiCoder uses the HeapSearch algorithm, which is a bottom-up enumerative search algorithm [8] outputting programs following the likelihood order of the prediction model. It has been shown superior to other enumeration methods (such as A* or Beam search) and easily deployed across parallel compute environments. The candidate programs use the results of the `SPARQL` queries on the knowledge graph to be run on the examples.

4 Evaluation

We perform experiments to answer the following questions:

(Q1) Which milestones as described in Sect. 2.2 can be achieved with our algorithm?
(Q2) How does WikiCoder compare to GPT-3 and Codex, the algorithm powering Copilot based on very large language models?

(Q3) Can a knowledge-powered program synthesis tool operate at scale for classical purely syntactic tasks?

4.1 Environment

The dataset along with the code for the experiments are available publicly on GitHub[2]. WikiCoder is written in Python, it leverages a prediction model written in PyTorch. We favoured simplicity over performance as well as clear separation into independent components as described above.

Benchmark Suite. Since there are no existing datasets to test the new aspects introduced in Sect. 2.2, we created one ourselves. The dataset is comprised of 46 tasks in the spirit of the FlashFill dataset [9]. Some of the tasks are inspired or extracted from [22], they are tagged as such in our code. Each task requires external knowledge to be solved and is labelled with 3 metadata:

- preprocessing for entity extraction: 0 if the inputs are already the sought entities, 1 if a syntactical program is enough to extract them, and 2 otherwise;
- complexity of the relationships between entities : 0 when no relation from the knowledge graph is needed, 1 for simple (single edge in the knowledge graph), and 2 for composite;
- postprocessing on external knowledge: 0 if the knowledge is used without postprocessing, and 1 otherwise.

This induces 8 categories, we provide at least 4 tasks for each category.

Knowledge Graph. For simplicity and reproducibility, we use a custom-made small knowledge graph for the experiments. We provide SPARQL queries to construct the knowledge graph.

Experimental Setup. All our experiments were performed on a consumer laptop (MSI GF65 Thin 9SE) with an intel i7-9750H CPU working up to 2.60 GHz on a single thread. The operating system is Manjaro 21.3 with linux kernel 5.17.5-1. The code is written in Python 3.8.10. The framework used for the SPARQL database is BlazeGraph 2.1.6 The code made available includes all the details to reproduce the experiments, it also includes the exact version of the Python libraries used.

Prediction Model. In our prediction model we embed the examples with a one hot encoding. Since the number of examples can change from one task to another, they are fed into a single layer RNN to build an intermediate representation of fixed shape. Then it is followed by 4 linear layers with ReLU activation functions, except for the last layer where no activation function is used. The final tensor is split, and a log softmax operator is applied for each non-terminal of the grammar, thus producing a probabilistic context free grammar in log space.

[2] https://github.com/nathanael-fijalkow/ProgSynth.

The prediction model is trained on a dataset of 2500 tasks for 2 epochs with a batch size of 16 tasks. The tasks were generated with the following process: a program was sampled randomly from a uniform probabilistic context free grammar, then inputs are randomly generated and run on the program. We used the Adam optimiser with a cross entropy loss to maximise the probability of the solution program being generated from the grammar.

4.2 Results on the New Dataset

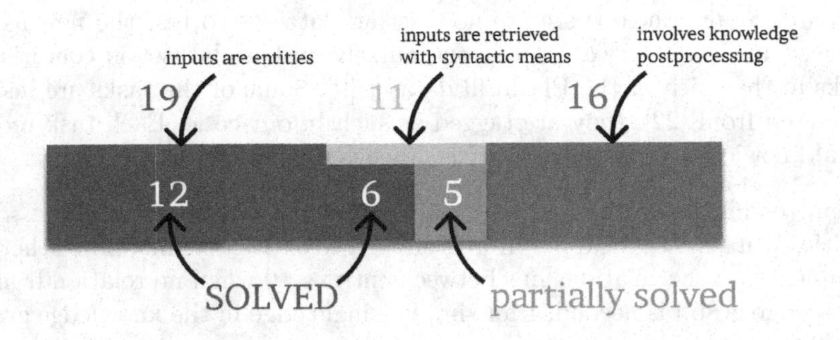

Fig. 5. WikiCoder performance on our new dataset

WikiCoder solves 18 out of 46 tasks with a timeout of 60 s. Let us make more sense of this number:

- Since our algorithm does not perform postprocessing on knowledge, none of the 16 tasks involving knowledge postprocessing were solved. Thus only 30 tasks are within reach of our algorithm.
- Among the 30 tasks, for 19 of them the inputs are directly the sought entities. WikiCoder solves 12 out of these 19 tasks, most failed tasked are not solvable with our DSL.
- In the remaining 11 tasks, the entities can be retrieved using syntactic manipulations. WikiCoder solves 6 out of these 11 tasks.
- Digging deeper for the last case, the fault lies in all remaining 5 cases with the preprocessing algorithm, which fails to correctly decompose the task and formulate the appropriate `SPARQL` queries. However, when provided with the right decomposition, WikiCoder solves all 5 tasks.

The results can be visualised on Fig. 5. The takeaway message is: WikiCoder solves almost all tasks which does not involve knowledge postprocessing, and when it does not the main issue is entities extraction with the preprocessing algorithm.

4.3 Comparison with Codex and GPT-3

It is not easy to perform a fair comparison with Codex and GPT-3 as they solve different problems. There are two ways these models can solve programming by example tasks. The experiments were performed on OpenAI's beta using the Da Vinci models (the most capable) for GPT-3 and Codex, with default parameters.

Using Codex: as *code completion*, provided with examples as docstring. This makes the problem very hard for Codex for two reasons: first, it was trained to parse specification in natural language, possibly enriched with examples. Providing only examples is a very partial specification which Codex is not used to. Second, Codex does not use a domain specific language, but rather general purpose programming languages such as Python and C. This implies that the program might be wrong in different ways: incorrect syntax, typing issues, or compilation errors. For the reasons above, in this infavourable comparison Codex solves only the 5 easiest tasks.

There are two ways of using GPT-3: *query answering* or *code completion*. For query answering, we feed GPT-3 with all but the last examples, and ask it to give the output for the input of the last example:

```
Query:
    f("France") = "I live in France"
    f("Germany") = "I live in Germany"
    f("Poland") = "I live in Poland"
    f("New Zealand") = ?
Output:
    f("New Zealand") = "I live in New Zealand"
```

The weakness of this scenario is that GPT-3 does not output a program: it only answers queries. One consequence is that the correctness test is very weak: getting the right answer on a single query does not guarantee that it would on any input. Worse, not outputting a program means that the whole process acts as a black-box, giving up on the advantages of our framework (see related work section) and program synthesis in general.

Using GPT-3/ChatGPT as *code completion*: the prompt is to generate a Python function that satisfies the following examples. This fixes the weakness of not outputting a program with query answering. However, on tasks that require a knowledge graph, the answer is either a succession of if statements or using a dictionnary which is semantically equivalent to the if statements. In some sense, the system having only partial knowledge, it does not attempt to generalise beyond the given examples.

GPT-3 performs very well in the query answering setting, solving 31 tasks out of 46. Only 2 tasks were solved by WikiCoder and not by GPT-3, and conversely 11 by GPT-3 and not by WikiCoder. GPT-3 only solves 3 tasks involving knowledge postprocessing: in this sense, WikiCoder and GPT-3 suffer from the same limitations, they both struggle with knowledge postprocessing.

4.4 Results on FlashFill

To show that WikiCoder operates at scale on classical program synthesis tasks, we test it against the classical and established FlashFill dataset. The results are shown on Fig. 6 (displaying cumulated time), WikiCoder solves 70 out of 101 tasks with a timeout of 60 s per task, on par with state of the art general purpose program synthesis tools. Only 85 tasks can be solved with our DSL.

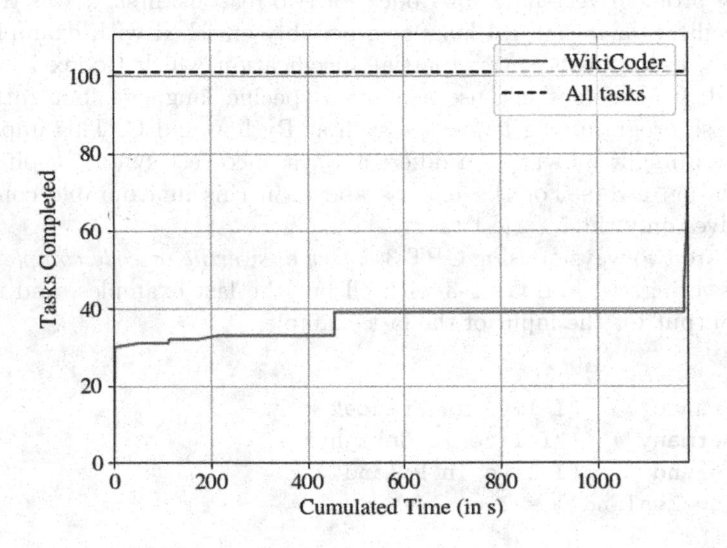

Fig. 6. WikiCoder solves 70% of FlashFill tasks

4.5 Limitations

To show the limitations of our approach, we discuss some counterexamples.

Entities Extraction. Our preprocessing algorithm looks for the longest shared pattern of length at least 2 in the examples. Let us consider the following example:

```
f("France") = "I live in France"
f("Germany") = "I live in Germany"
f("Poland") = "I live in Poland"
```

The longest pattern is naturally "I live in", however another pattern occurs, unexpectedly: "an" appears in each country's name. When using our preprocessing algorithm this leads to a wrong sketch. Adding a fourth example would remove this ambiguity if the country name does not include "an". A related example would be if all examples include year dates in the same millenium, say all of the form "20XX": then "20" appears in each example, but it should not be separated from XX.

Knowledge Postprocessing. None of the tasks involving knowledge postprocessing were solved. Indeed, if the entity to be used is not present in the example, it is very hard to guess which one it is. Natural candidates are entities in the neigbourhood of the starting entity. Our approaches to this challenge have proved inefficient.

4.6 Examples of Programs

To show that the programs generated are clear and interpretable by humans, we show a few of them translated into Python equivalents. Here is the first example where two relations are needed:

```
// Examples:
   f("France") = "French, capital:Paris"
   f("Germany") = "German, capital:Berlin"
   f("China") = "Chinese, capital:Beijing"
   f("New Zealand") = "New Zealander, capital:Wellington"

// Generated program:
def f(x: str) -> str:
   a = label(follow_edges_from(x, "demonym"))
   b = ", capital:"
   c = label(follow_edges_from(x, "isCapitalOf"))
   return a + b + c
```

We present another example where a relation at distance 2 is needed:

```
// Examples:
   f("Paris") = "The phone country code is 33"
   f("Berlin") = "The phone country code is 49"
   f("Detroit") = "The phone country code is 1"
   f("Chihuahua") = "The phone country code is 52"

// Generated program:
def f(x: str) -> str:
   a = "The phone country code is "
   b = label(follow_edges_from(x, "CityOf", "phoneCode"))
   return a + b
```

5 Discussion

5.1 Related Work

The closest related work is [22]: the tool FlashGPT3 solves knowledge-powered program synthesis with a completely different approach: instead of querying a knowledge graph, FlashGPT3 uses a large language model (GPT3 [3]) to get external knowledge. FlashGPT3 is shown superior to both program synthesis tools and GPT3 taken in isolation, and achieves impressive results in different domains. There are three advantages of using a knowledge graph over a large language model:

- **Reliability**: the synthesized program is only as reliable as the knowledge source that it uses. To illustrate this point, let us consider the task
  `f("Paris") = "France"`
  `f("Berlin") = "Germany"`
  GPT3 needs very few examples to predict that
  `f("Washington") = "United States"`
  but it might fail on more exotic inputs. On the other hand, querying a knowledge graph results in setting `f = CountryOf` implying that the program will correctly reproduce the knowledge from the graph. Although GPT3 has achieved extraordinary results in query answering, it is still more satisfactory to rely on an established and potentially certified source of knowledge such as Wikipedia.
- **Explainability**: the constructed program has good explainability properties: the exact knowledge source it uses can be traced back to the knowledge graph and therefore effectively verified.
- **Adaptability**: knowledge-powered program synthesis can be deployed with any knowledge graph, possibly collecting specialised or private information which a largue language model would not know about.

Other Knowledge-Powered Program Synthesis Tools. Most program synthesis tools work at a purely syntactical level. However, some included limited level of semantic capabilities for domain-specific tasks: for instance Transform-data-by-example (TDE) uses functions from code bases and web forms to allow semantic operations in inductive synthesis [11], and APIs were used for data transformations [2].

Knowledge Graphs. A growing research community focuses on creating, maintaining, and querying knowledge graphs. The most related problems to our setting are query by example, where the goal is either to construct a query from a set of examples [15,18], and entity set expansion, aiming at expanding a small set of examples into a more complete set of entities having common traits [25].

5.2 Contributions and Outlook

We have introduced knowledge-powered program synthesis, extending program synthesis by allowing programs to rely on knowledge graphs. We described a number of milestones for this exciting and widely unexplored research objective, and proposed a dataset to evaluate the progress in this direction. We constructed an algorithm and implemented a general-purpose knowledge-powered program synthesis tool WikiCoder that solves tasks previously unsolvable. Our tool can only address about one third of the dataset; we believe that solving the whole dataset would be an important step forward towards deploying program synthesis in the real world.

The most natural continuation of this work is to use very large language models for knowledge-powered program synthesis. As discussed above, how can we retain the properties of our framework with knowledge graphs: reliability, explainability, and adaptability, while leveraging the power of very large language models?

References

1. Balog, M., Gaunt, A.L., Brockschmidt, M., Nowozin, S., Tarlow, D.: Deepcoder: learning to write programs. In: International Conference on Learning Representations, ICLR (2017)
2. Bhupatiraju, S., Singh, R., Mohamed, A., Kohli, P.: Deep API programmer: learning to program with APIs. CoRR, abs/1704.04327 (2017)
3. Brown, T.B., et al.: Language models are few-shot learners. In: Annual Conference on Neural Information Processing Systems, NeurIPS (2020)
4. Chen, M., et al.: Evaluating large language models trained on code. CoRR, abs/2107.03374 (2021)
5. Chen, X., Liu, C., Song, D.: Execution-guided neural program synthesis. In: International Conference on Learning Representations, ICLR (2018)
6. Devlin, J., Uesato, J., Bhupatiraju, S., Singh, R., Mohamed, A., Kohli, P.: Robustfill: neural program learning under noisy I/O. In: International Conference on Machine Learning, ICML, vol. 70 of Proceedings of Machine Learning Research (2017)
7. Ellis, K., et al.: Dreamcoder: bootstrapping inductive program synthesis with wake-sleep library learning. In: International Conference on Programming Language Design and Implementation, PLDI (2021)
8. Fijalkow, N., Lagarde, G., Matricon, T., Ellis,v K., Ohlmann, P., Potta, A.: Scaling neural program synthesis with distribution-based search. In: International Conference on Artificial Intelligence. AAAI (2022)
9. Gulwani, S.: Automating string processing in spreadsheets using input-output examples. In: ACM SIGPLAN-SIGACT Symposium on Principles of Programming Languages, POPL (2011)
10. Gulwani, S., Polozov, O., Singh, R.: Program synthesis. Found. Trends Program. Lang. **4**(1–2), 1–19 (2017)
11. He, Y., Chu, X., Ganjam, K., Zheng, Y., Narasayya, V.R., Chaudhuri, S.: Transform-data-by-example (TDE): an extensible search engine for data transformations. Proc. VLDB Endow. **11**(10), 1165–1177 (2018)
12. Hoffart, J., Suchanek, F.M., Berberich, K., Weikum, G.: YAGO2: a spatially and temporally enhanced knowledge base from wikipedia. Artif. Intell. **194**, 28–61 (2013)
13. Hogan, A.: Knowledge graphs: a guided tour (invited paper). In: Bourgaux, C., Ozaki, A., Peñaloza, R. (eds.) International Research School in Artificial Intelligence in Bergen, AIB 2022, University of Bergen, Norway, 7–11 June 2022, vol. 99 of OASIcs, pp. 1:1–1:21. Schloss Dagstuhl - Leibniz-Zentrum für Informatik (2022)
14. Hogan, A., et al.: Knowledge graphs. In: Synthesis Lectures on Data, Semantics, and Knowledge. Morgan & Claypool Publishers (2021)

15. Jayaram, N., Khan, A., Li, C., Yan, X., Elmasri, R.: Querying knowledge graphs by example entity tuples. In: 32nd IEEE International Conference on Data Engineering, ICDE 2016, Helsinki, Finland, 16–20 May 2016, pp. 1494–1495. IEEE Computer Society (2016)
16. Kalyan, A., Mohta, A., Polozov, O., Batra, D., Jain, P., Gulwani, S.: Neural-guided deductive search for real-time program synthesis from examples. In: International Conference on Learning Representations, ICLR (2018)
17. Lee, W., Heo, K., Alur, R., Naik, M.: Accelerating search-based program synthesis using learned probabilistic models. In: ACM SIGPLAN Conference on Programming Language Design and Implementation, PLDI (2018)
18. Metzger, S., Schenkel, R., Sydow, M.: QBEES: query-by-example entity search in semantic knowledge graphs based on maximal aspects, diversity-awareness and relaxation. J. Intell. Inf. Syst. **49**(3), 333–366 (2017). https://doi.org/10.1007/s10844-017-0443-x
19. Polosukhin, I., Skidanov, A.: Neural program search: solving programming tasks from description and examples. In: International Conference on Learning Representations, ICLR (2018)
20. Shi, K., Bieber, D., Singh, R.: Tf-coder: program synthesis for tensor manipulations. ACM Trans. Program. Lang. Syst. **44**(2), 10:1–10:36 (2022)
21. Pellissier Tanon, T., Weikum, G., Suchanek, F.: YAGO 4: a reason-able knowledge base. In: Harth, A., et al. (eds.) ESWC 2020. LNCS, vol. 12123, pp. 583–596. Springer, Cham (2020). https://doi.org/10.1007/978-3-030-49461-2_34
22. Verbruggen, G., Le, V., Gulwani, S.: Semantic programming by example with pre-trained models. Proc. ACM Program. Lang. **5**(OOPSLA), 1–25 (2021)
23. Wang, C., Cheung, A., Bodík, R.: Synthesizing highly expressive SQL queries from input-output examples. In: Cohen, A., Vechev, M.T. (eds.) Conference on Programming Language Design and Implementation, PLDI, pp. 452–466. ACM (2017)
24. Zhang, L., et al.: Leveraging constraint logic programming for neural guided program synthesis. In: International Conference on Learning Representations, ICLR (2018)
25. Zheng, Y., Shi, C., Cao, X., Li, X., Wu, B.: A meta path based method for entity set expansion in knowledge graph. IEEE Trans. Big Data **8**(3), 616–629 (2022)
26. Zhou, X., Bodík, R., Cheung, A., Wang, C.: Synthesizing analytical SQL queries from computation demonstration. In: Jhala, R., Dillig, I. (eds.) International Conference on Programming Language Design and Implementation, PLDI, pp. 168–182. ACM (2022)
27. Zohar, A., Wolf, L.: Automatic program synthesis of long programs with a learned garbage collector. In: Neural Information Processing Systems, NeurIPS (2018)

Provable Correct and Adaptive Simplex Architecture for Bounded-Liveness Properties

Benedikt Maderbacher[1(✉)], Stefan Schupp[2], Ezio Bartocci[2], Roderick Bloem[1], Dejan Ničković[3], and Bettina Könighofer[1]

[1] Graz University of Technology, Graz, Austria
{benedikt.maderbacher,roderick.bloem,bettina.koenighofer}@iaik.tugraz.at
[2] TU Wien, Vienna, Austria
{stefan.schupp,ezio.bartocci}@tuwien.ac.at
[3] AIT Austrian Institute of Technology, Vienna, Austria
dejan.nickovic@ait.ac.at

Abstract. We propose an approach to synthesize Simplex architectures that are *provably correct* for a rich class of temporal specifications, and are *high-performant* by optimizing for the time the advanced controller is active. We achieve provable correctness by performing a static verification of the baseline controller. The result of this verification is a set of states which is proven to be safe, called the *recoverable region*. During runtime, our Simplex architecture *adapts* towards a running advanced controller by exploiting proof-on-demand techniques. Verification of hybrid systems is often overly conservative, resulting in over-conservative recoverable regions that cause unnecessary switches to the baseline controller. To avoid these switches, we invoke targeted reachability queries to extend the recoverable region at runtime.

Our offline and online verification relies upon reachability analysis, since it allows observation-based extension of the known recoverable region. However, detecting fix-points for bounded liveness properties is a challenging task for most hybrid system reachability analysis tools. We present several optimizations for efficient fix-point computations that we implemented in the state-of-the-art tool HyPro that allowed us to automatically synthesize verified and performant Simplex architectures for advanced case studies, like safe autonomous driving on a race track.

1 Introduction

With the unprecedented amount of available computational power and the proliferation of artificial intelligence, modern control applications are becoming

This project has received funding from the European Union's Horizon 2020 research and innovation programme under grant agreement № 956123 - FOCETA, the Austrian FWF project ZK-35, the austrian research promotion agency FFG projects ADVANCED (№ 874044) and FATE (№ 894789), the Graz University of Technology LEAD Project *Dependable Internet of Things in Adverse Environments*, and the State Government of Styria, Austria – Department Zukunftsfonds Steiermark.

G. Caltais and C. Schilling (Eds.): SPIN 2023, LNCS 13872, pp. 141–160, 2023.
https://doi.org/10.1007/978-3-031-32157-3_8

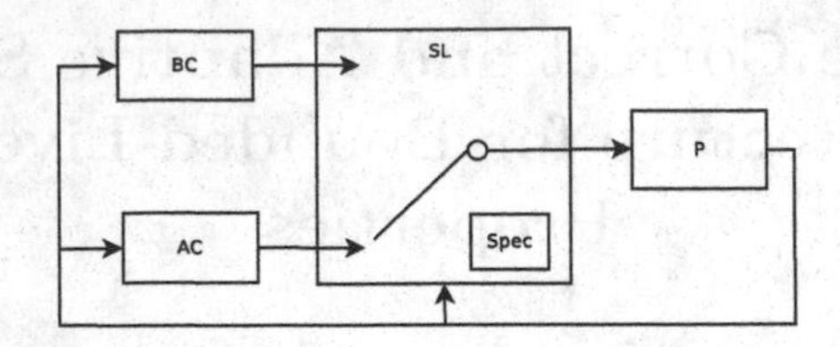

Fig. 1. Schematic of a Simplex architecture

increasingly autonomous. The increasing adoption of the DevOps paradigm by the cyber-physical systems (CPSs) community facilitates development of the control systems beyond their deployment – the observation and collection of data during system operation allows control systems to evolve and improve over time.

Developing trusted advanced controllers has therefore become a major challenge in safety-critical domains. Since the high complexity of advanced controllers renders formal verification infeasible, runtime assurance methods [26], which ensure safety by monitoring and altering the execution of the controller, gain more and more importance.

Simplex Architecture. The Simplex architecture [7,25] is a well-established runtime assurance architecture, originally proposed for reliable upgrades in a running control system. Let the *plant* P be the physical system of hybrid nature (the system's components exhibit mixed discrete-continuous behavior), and φ be the safety specification. The Simplex architecture that ensures that P works within the specification φ, consists of three components, as illustrated in Fig. 1:

1. The *baseline controller* (BC) is a formally verified controller with respect to the given specification φ. Therefore, it is proven that the baseline controller provides control inputs to the plant in such a way that φ is satisfied. This holds under the assumption that the plant is initially in a state from which the baseline controller is able to satisfy φ. We call the set of plant's states, from which the baseline controller guarantees safe operation according to φ, the *recoverable region.*
2. The *advanced controller* (AC), a highly efficient and complex controller that might incorporate deep neural networks. However, due to its complexity, the advanced controller cannot be formally verified and therefore we have no guarantees whether it always satisfies φ.
3. The *switching logic* (SL) monitors the execution of the advanced controller and hands the control to the baseline controller if the plant would otherwise leave the recoverable region where the baseline controller guarantees φ.

The Simplex architecture *guarantees safety* by ensuring that the plant operates within the formally proven, recoverable region of the baseline controller and facilitates *high performance* by enabling the advanced controller maxima freedom and only restricting its operation to the recoverable region.

Challenges. While conceptually simple, the implementation of a Simplex architecture is challenging. To guarantee safety, it is required to formally verify the baseline controller. Hybrid automata [2] are a common model for accurately describing composite systems that combine discrete computational and continuous processes, therefore hybrid automata are the desired model for baseline controllers. However, the verification of hybrid automata is typically the bottleneck due to the inherent limits of exhaustive verification tools. To obviate formal verification, it is common practice to make assumptions about the correctness of a given baseline controller and its recoverable region. By doing so, one loses all correctness guarantees the Simplex architecture should provide.

The second challenge is to implement a Simplex architecture that allows high performance while being formally verified. In case the formal verification of the baseline controller was successful, the accumulation of over-approximation errors during the static verification phase often results in an over-conservative recoverable region. Therefore, the switching logic gives control to the baseline controller much more often than necessary, causing unnecessary drops in performance.

Problem Statement. We start from a given plant and a baseline controller, both modeled as hybrid automata, and a safety specification expressed in signal temporal logic (STL) [14], which includes bounded-liveness requirements. The problem is to synthesize Simplex architectures that are provably correct for the given specification for unbounded time, and are high-performant by optimizing for the time the advanced controller is active.

Our Approach. We compute at design time an initial recoverable region as the set of states reachable by the baseline controller from an initial region. During the reachability analysis, we check that the specification is satisfied for all reachable states from the baseline controller starting from its initial states. Using a fixpoint detection on the set of reachable states allows us to guarantee safety with respect to the given specification on infinite time.

We then optimize, during the execution, the performance of the operating Simplex architecture incrementally by adapting towards the running advanced controller. Whenever the advanced controller proposes an output that would cause the plant to leave the current recoverable region, we perform two steps: we switch to the baseline controller, and we try to enlarge the recoverable region. To enlarge the recoverable region, we collect *suspicious states* that would have been reached in case the last command of the advanced controller would have been executed. Next, we analyze the behavior of the baseline controller on all suspicious states. If we are able to prove that the baseline controller ensures safety for a suspicious state, we add this state and all states reachable from it to the recoverable region. Next time the advanced controller tries to enter such a state, the switching logic will not interfere.

Contributions. Our main results can be summarized as follows:

- We propose a workflow to synthesize *provable correct* Simplex architectures that *dynamically adapt* to a running advanced controller.
- Our proposed methodology enables a lightweight reachability analysis to show safety of suspicious states on demand. To this end, we employ flowpipe

construction based reachability analysis as an inductive approach which allows to compute sets of reachable states for given initial configurations. This enables local updates of the known recoverable region in a light-weight fashion based on current observations of the system in an iterative way enabling us to involve information obtained at runtime. We aim to incorporate already established knowledge about recoverable regions into the verification approach to further improve its scalability.
- To the best of our knowledge, we are the first that synthesize Simplex architectures for bounded-liveness specifications.
- We present several optimizations for efficient fixpoint computations. In particular, we introduce novel data structures for faster lookup as well as set-theoretic and symbolic approaches to improve fixpoint detection. We implemented these optimizations in the hybrid system reachability analysis tool HyPro [24]. Only with our optimizations, we were able to verify hybrid automata against bounded liveness properties in HyPro.
- We present a detailed case study on safe driving on a racing track and the effects of optimizing the Simplex architecture during runtime.

Related Work. Original works on the Simplex architecture [7,25,26], as well as many recent work [12,19,27] assume to have a verified baseline controller and a correct switching logic given. Under these assumptions, the papers guarantee safe operation of the advanced controller. However, these assumptions are very strong and the works ignore the challenges and implied limitations that need to be addressed in order to get a verified baseline controller and a switching logic that is guaranteed to switch at the correct moment. The reason for many works to leave out these steps is that a general safety statement for *unbounded time* for the baseline controller is required (i.e., a fixpoint in the analysis). Depending on the utilized method, fixpoints in the analysis cannot always be found, as some verification methods tend to have bad convergence due to accumulating errors.

Recent works that verify the baseline controller deploy standard methods to verify hybrid systems such as barrier certificates [17,18,29], and using forward or backward reachability analysis [4]. Our method is independent of the concrete approach that is used for offline verification. In general, for methods based on flowpipe construction for forward or backward reachability analysis, there exists a trade-off between accuracy and complexity: using simple shapes to over-approximate the reachability tubes results in overly-conservative recoverable regions, while using too complex shapes requires difficult computations to check for a fixpoint. By using the concept of proof on demand, we allow simple shapes for the reachability analysis but amend the problem of an over-conservative recoverable region by enlarging the region on demand.

While several works study provable correct Simplex architectures, there is only little work on how to create high-performant Simplex architectures. Similar to our approach, the work in [13] uses online computations to increase performance. While our approach adapts to a given advanced controller and therefore the number of online proofs reduces during exploitation, the approach in [13] performs the same online computations repeatedly.

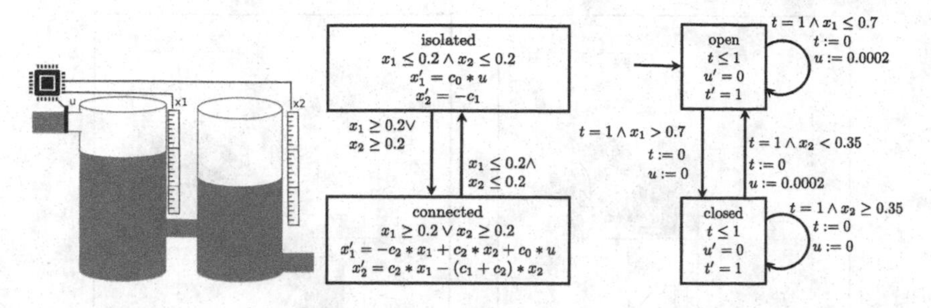

Fig. 2. Left: Illustration of the plant. Middle: Hybrid automaton $\mathcal{H}_P$ modeling the plant. Right: Hybrid automaton $\mathcal{H}_{BC}$ implementing the baseline controller.

Furthermore, to the best of our knowledge, no work considered temporal logic properties beyond safety invariants to be enforced by a Simplex architecture. Instead, our work allows us to specify bounded reachability and bounded liveness properties and conjunctions of these properties.

Runtime assurance covers a wide range of techniques and has several application areas, for example enforcing safety in robotics [15] or in machine learning [28]. Runtime enforcers, often called shields, directly alter the output of the controller during runtime to enforce safety. In the discrete setting, such enforcers can be automatically computed from a model of the plant's dynamics and the specification using techniques from model checking and game theory [1]. In the continuous domain, inductive safety invariants such as a control Lyapunov functions [21] or control barrier functions [20] are used to synthesize runtime enforcers.

2 Illustrative Example

As an illustrative example, we use a well-known textbook example of a coupled water tank system [5], as illustrated in Fig. 2 (left). Using this example, we will outline how we construct the Simplex architecture from a given safety specification in STL and a given baseline controller in form of a hybrid automaton.

Plant P. The plant consists of two tanks that are connected by a pipe that is located at a height of 0.2 m from the floor. The left tank has an inflow that can be adjusted by a controller. The right tank has an outflow pipe that constantly drains water. The plant with the two connected water tanks can be modeled by the hybrid automaton $\mathcal{H}_P$ given in Fig. 2 (middle). The automaton has two state variables x_1 and x_2, corresponding the level of water in the left and right tank respectively, and one control dimension $u \in [0, 0.0005]$ influencing how much water is added to the left tank. The two states reflect the two modes with different dynamics. If the water in any of the tanks is higher than 0.2 m water can flow through the connecting pipe and the levels in the two tanks equalize. Otherwise, the tanks are isolated and evolve only according to their own dynamics.

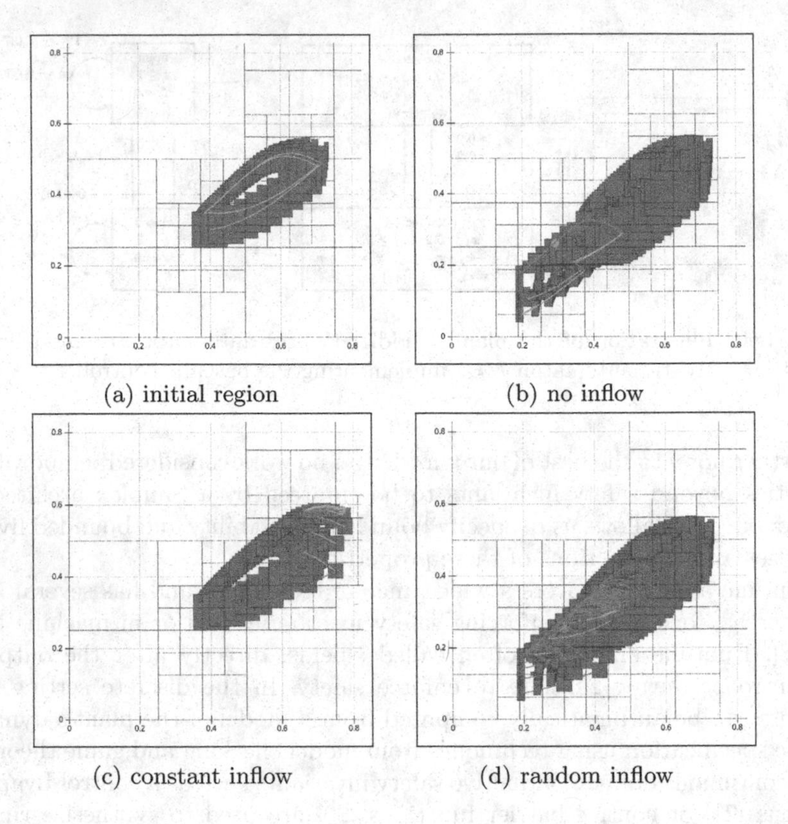

(a) initial region (b) no inflow

(c) constant inflow (d) random inflow

Fig. 3. Visualisation of the initial recoverable region and adapted recoverable regions of the different advanced controllers for variables x_1, x_2. (Color figure online)

Safety Specification φ**.** The safety specification $\varphi = \varphi_1 \wedge \varphi_2$ requires that the following two properties φ_1 and φ_2 are satisfied:

1. The water tanks may not be filled beyond their maximum filling height of 0.8 m. This property can be expressed in STL via $\varphi_1 = \mathbf{G}(x_1 \leq 0.8 \wedge x_2 \leq 0.8)$.
2. If the water level of the right tanks falls below 0.12 m it has to be filled up to at least 0.3 m within the next 30 time units. This is written in STL as $\varphi_2 = \mathbf{G}(x_2 \leq 0.12 \rightarrow \mathbf{F}_{[0,30]}(x_2 \geq 0.3))$.

A safety specification φ in STL can be transformed into a hybrid automaton $\mathcal{H}_\varphi$. We discuss this transformation in the next section.

Baseline Controller. We use the baseline controller that is given by the hybrid automaton $\mathcal{H}_{BC}$ in Fig. 2. $\mathcal{H}_{BC}$ consists of two locations *open* and *closed.* At the end of a cycle ($c = 1$) the controller observes x_1 and x_2 to determine whether to stay in its current location or switch to the other one. $\mathcal{H}_{BC}$ remains in *open* as long as $x_1 \leq 0.7$ and in *closed* as long as $x_2 \geq 0.35$. After every transition, the value of u is set to either 0 or 0.0002 depending on the target location.

Static Verification for a Conservative Simplex Architecture. In the first step, we verify the baseline controller $\mathcal{H}_{BC}$ acting within the plant $\mathcal{H}_P$ with respect to the specification $\mathcal{H}_\varphi$ for the initial states of the plant $x_1 \in [0.35, 0.45]$ and $x_2 \in [0.25, 0.35]$. To compute the initial recoverable region, we compute the fixpoint of reachable states while checking that no bad state defined by $\mathcal{H}_\varphi$ is contained in the reachable states. We verified the baseline controller using the reachability analysis tool HyPro [24]. Figure 3(a) illustrates the initial recoverable region. The blue squares depict the initially known recoverable region for each control cycle. The orange trajectory shows the behavior of the plant when controlled by the baseline controller.

Adapting to the Advanced Controller via Proof on Demand. In the second step, we run the Simplex architecture for 500 control cycles and evolve the recoverable region during runtime. Figure 3(b)-(d) shows the growth of the recoverable region over time for different advanced controllers: the first advanced controller (Fig. 3(b)) sets u to 0 constantly, the second advanced controller (Fig. 3(c)) sets u to 0.0004 constantly, and the third advanced controller (Fig. 3(d)) picks $u \in [0.0001, 0.0005]$ equally distributed with a 10 % probability, and sets $u = 0$ otherwise. In each figure, the green trajectory shows the behavior of the plant when being controlled by the corresponding advanced controller. The plots only show the known recoverable region at the end of each control cycle. The points where the trajectory exits the blue region these points are between control cycles and have been verified safe, but they are not stored for efficiency reasons. In these experiments, we check the safety of states outside of the recoverable region on the fly and immediately add them to the recoverable region if they are safe. The switching logic only switches to the baseline controller, if the advanced controller would visit unsafe states that cannot be added to the recoverable region. Note, that these proofs on demand can also be performed in the background.

3 Background

3.1 Reachability Analysis of Hybrid Systems

We use *hybrid automata* as a formal model for hybrid systems.

Definition 1 (Hybrid automata [2]**).** *A hybrid automaton* $\mathcal{H} = (Loc, Lab, Edge, Var, Init, Inv, Flow, Jump)$ *consists of a finite set of locations* $Loc = \{\ell_1, \ldots, \ell_n\}$*, finite set of labels* Lab*, which synchronize and coordinate state changes between automata, a finite set of jumps* $Edge \subseteq Loc \times Lab \times Loc$*, that allow to realize location changes, a finite set of variables* $Var = \{x_1, \ldots, x_d\}$*, a set of states* Inv *called invariant, which restricts the values* ν *for each location, a set of initial states* $Init \subseteq Inv$*, a flow relation* $Flow$ *where* $Flow(\ell) \subseteq \mathbb{R}^{Var} \times \mathbb{R}^d$*, which determines for each state* (ℓ, ν) *the set of possible derivatives* $\dot{Var}$*, a jump relation* $Jump$ *where* $Jump(e) \subseteq \mathbb{R}^d \times \mathbb{R}^{d'}$ *defines for each jump* $e \in Edge$ *the set of possible successors* ν' *of* ν*.*

A state $\sigma = (\ell, \nu)$ of $\mathcal{H}$ consists of a location ℓ and a valuation $\nu \in \mathbb{R}^d$ for each variable, $S = Loc \times \mathbb{R}^d$ denotes the set of all states.

Every behavior of $\mathcal{H}$ must start in one of the initial states $Init \subseteq Inv$. Jump relations are typically described by a guard set $G \subseteq \mathbb{R}^d$ and an assignment (or reset) $\nu' = r(\nu)$ as $Jump(e) = \{(\nu, \nu') \mid \nu \in G \wedge \nu' = r(\nu)\}$. For simplicity, we restrict ourselves to the class of *linear hybrid automata*, i.e., the dynamics ($Flow$) are described by systems of linear ordinary differential equations and guards (G), invariant conditions (Inv), and sets of initial variable valuations are described by linear constraints. Resets on discrete jumps are given as affine transformations. In this work, we use composition of hybrid automata as defined in [10] with label-synchronization on discrete jumps and shared variables.

A *path* $\pi = \sigma_1 \rightarrow_\tau \sigma_2 \rightarrow_e \rightarrow \cdots$ in $\mathcal{H}$ is an ordered sequence of states σ_i connected by *time transitions* $\rightarrow_\tau$ of length τ and *discrete jumps* $\rightarrow_e, e \in Edge$. Time transitions follow the flow relation while discrete jumps follow the edge and jump relations, we refer to [10] for a formal definition of the semantics. Paths naturally extend to *sets of paths* which collect paths with the same sequence of locations but different variable valuations.

The Reachability Problem in Hybrid Automata. A state $\sigma_i = (\ell, \nu)$ of $\mathcal{H}$ is called *reachable*, if there is a path π leading to it with $\sigma_1 \in Init$. The reachability problem for hybrid automata tries to answer whether a given set of states $S_{bad} \subseteq S$ is reachable. Since the reachability problem is in general undecidable [11], current approaches often compute over-approximations of the sets of reachable states for *bounded reachability*. Note that reachability analysis follows all execution branches, i.e., does not resolve any non-determinism induced by discrete jumps in the model. That means, that computing alternatingly time- and jump-successor states may yield a tree-shaped structure (nodes contain time-successors, the parent-child relation reflects discrete jumps, see also [22]) which covers all possible executions.

Flowpipe Construction for Reachability Analysis. For a given hybrid automaton $\mathcal{H}$, flowpipe construction (see e.g., [6]) computes a set of convex sets

$$R = reach^{\mathcal{H}}_{\leq\alpha}(\sigma),$$

which are guaranteed to cover all trajectories of bounded length α that are reachable from a set of states σ. We use $reach^{\mathcal{H}}_{=\alpha}(\sigma)$ to denote the set of states that are reached after exactly α time, and similarly $reach^{\mathcal{H}}_{\infty}(\sigma)$ to denote the set of states reachable for unbounded time.

The method over-approximates time-successor states by a sequence of sets (segments), referred to as *flowpipe*. Segments that satisfy a guard condition of an outgoing jump of the current location allow taking said jump leading to the next location. Note that non-determinism on discrete jumps may introduce branching i.e., it requires the computation of more than one flowpipe. The boundedness of the analysis is usually achieved by limiting the length of a flowpipe and the number of discrete jumps.

To compute the set of reachable states $reach^{\mathcal{H}}_{\infty}(\sigma)$ for unbounded time requires finding a fixpoint in the reachability analysis. For flowpipe-construction based techniques, finding fixpoints boils down to validating, whether a computed set of reachable states is fully contained in the set of previously computed state sets. As the approach accumulates over-approximation errors over time, it may happen that this statement cannot be validated [23]. In practice, researchers often check, whether the set obtained after a jump is contained in one of the already computed state sets.

Safety Verification via Reachability Analysis. Reachability analysis can be used to verify safety properties by checking that the reachable states do not contain any unsafe states. A system is (bounded-)safe if $R \cap S_{bad} = \emptyset$, otherwise the result is inconclusive. Unbounded safety results can only be obtained in case the method is able to detect a fixpoint for all possible trajectories in all possible execution branches.

3.2 Temporal Specification

We use STL [14], as the temporal specification language to express the safe behavior of our controllers. Let Θ be a set of terms of the form $f(R)$ where $R \subseteq S$ are subsets of variables and $f : \mathbb{R}^{|R|} \to \mathbb{R}$ are interpreted functions. The syntax of STL is given by the following grammar and we use standard semantics [14].

$$\varphi ::= \mathbf{true} \mid f(R) > k \mid \neg\varphi \mid \varphi_1 \vee \varphi_2 \mid \varphi_1 \mathbf{U}_I \varphi_2\,,$$

where $f(R)$ are terms in Θ, k is a constant in $\mathbb{Q}$ and I are intervals with bounds that are constants in $\mathbb{Q} \cup \{\infty\}$. We omit I when $I = [0, \infty)$. From the basic definition of STL, we can derive other standard operators as usual: conjunction $\varphi_1 \wedge \varphi_2$, implication $\varphi_1 \to \varphi_2$, eventually $\mathbf{F}_I\varphi$ and always $\mathbf{G}_I\varphi$.

In our approach, we use STL specifications to handle properties beyond simple invariants. More specifically, we support the following subset of STL specifications: *invariance* $\mathbf{G}(\varphi)$, *bounded reachability* $\mathbf{F}_{[0,t]}(\varphi)$, and *bounded liveness* $\mathbf{G}(\psi \to \mathbf{F}_{[0,t]}(\varphi))$ where φ and ψ are predicates over state variables and t is a time bound. We also allow assumptions about the environment such as the bounds on input variables.

STL specifications can be translated to hybrid automaton monitors. The translation is inspired by the templates used by Frehse et al. [8]. We adapt the original construction to facilitate fixpoint detection, by creating (mostly) deterministic monitors instead of universal ones. Figure 4 depicts the specification automata for the STL fragments considered in this work. A specification is violated when the *sink* location ℓ_{bad} is reached. Urgent transitions are encoded with location invariants and transition guards, such that no time may pass when an urgent transition is enabled. This is possible because our φ and ψ are half-plane constraints and we use the inverted guards as invariants. Conjunction of *invariance*, *bounded reachability*, and *bounded liveness* properties is enabled by parallel composition of monitor automata.

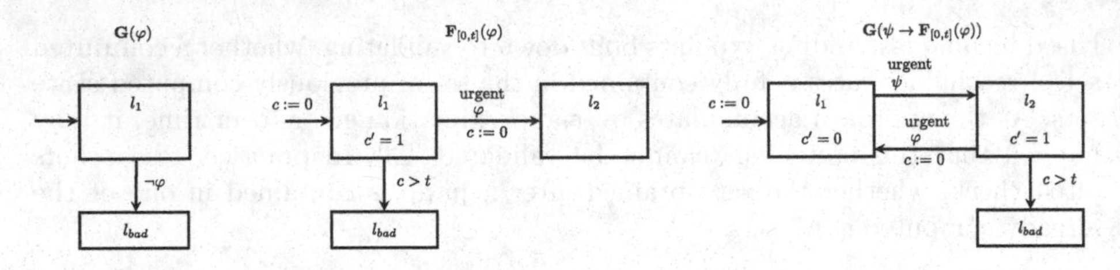

Fig. 4. Hybrid automata templates for the STL properties: *invariance* ($\mathbf{G}(\varphi)$), *bounded reachability* ($\mathbf{F}_{[0,t]}(\varphi)$), and *bounded liveness* ($\mathbf{G}(\psi \to \mathbf{F}_{[0,t]}(\varphi))$).

4 Verification of Adaptive Simplex Architectures

This section describes our workflow for obtaining a fully verified and adaptive Simplex architecture that is able to enforce the safety of a black box controller for temporal specifications. In this section, we first describe the setting and introduce the terminology in Sect. 4.1. Next, in Sect. 4.2, we describe how to verify a conservative known recoverable region and the techniques required to analyze systems for infinite time. Finally, in Sect. 4.3 we describe how to execute the Simplex architecture for temporal specifications and stateful baseline controllers, as well as how to incrementally extend the known recoverable region using proofs on demand.

4.1 Setting

We assume that we have a model of the plant, the baseline controller, and a bounded liveness specification. The plant model is given as a hybrid automaton $\mathcal{H}_P$ with variables $Var_{\mathcal{H}_P}$ and locations $Loc_{\mathcal{H}_P}$. We designate a subset $U \subseteq Var_{\mathcal{H}_P}$ of the variables as controller inputs with have constant dynamics and we use variables $X = Var_{\mathcal{H}_P}/U$ for the observable state of $\mathcal{H}_P$. We assume that the plant and its model $\mathcal{H}_P$ are deterministic, i.e., that for every observable variable valuation of the plant X there exists *exactly one* location ℓ in the model where $X \in Inv_{\mathcal{H}_P}$.

The baseline controller is a hybrid automaton $\mathcal{H}_{BC}$ equipped with a clock t that monitors the cycle time. $\mathcal{H}_{BC}$ is composed with $\mathcal{H}_P$ and can read the state $(\ell, \nu_X) \in Loc_{\mathcal{H}_P} \times \mathbb{R}^{|X|}$ of $\mathcal{H}_P$. At the end of each control cycle ($t = \delta$), $\mathcal{H}_{BC}$ sets the value of U via resets on synchronized jumps. An example of the two components is depicted in Fig. 2.

The specification of the system is given as an STL formula φ that is converted to a hybrid automaton $\mathcal{H}_\varphi$ as described in Sect. 3.2. The advanced controller is a black box that accesses the observable state (ℓ, ν_X) of the plant at the end of a control cycle and suggests an output, i.e., an assignment for each variable in U for the next control cycle.

Following classical terms, we partition the state space of $\mathcal{H}_P \times \mathcal{H}_{BC} \times \mathcal{H}_\varphi$ into the *unsafe region* S_{bad} (the specification has been violated) and the *safe region* S_{safe} (the specification holds). The region for which the baseline controller

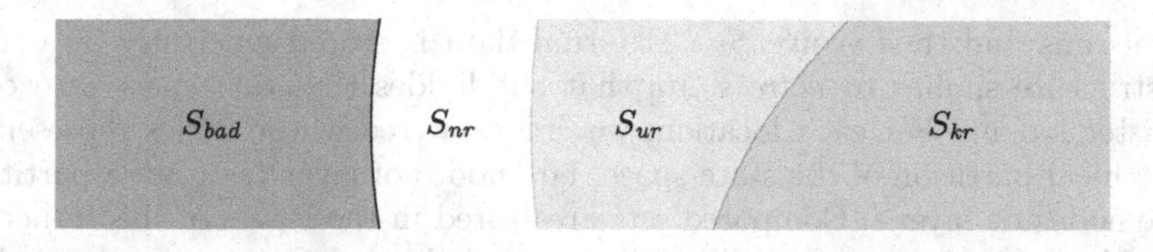

Fig. 5. Partitioning of the state space induced by a Simplex architecture.

satisfies the specification is referred to as the *recoverable region* $S_r \subseteq S_{safe}$, its complement is called the *non-recoverable region* S_{nr}.

The baseline controller is first verified for a small set of initial states. From verification, we obtain a subset of the recoverable region. In the following, we use the term *known recoverable region* S_{kr} for the states for which trust has been proven; the remainder of the recoverable region is the *unknown recoverable region* S_{ur}. The regions and their relations are illustrated in Fig. 5.

4.2 Static Verification of the Baseline Controller

We perform a reachability analysis based on a flowpipe construction. We detect fixpoints to verify that the baseline controller satisfy φ. The hybrid automaton $\mathcal{H} = \mathcal{H}_P \times \mathcal{H}_\varphi$ is the product of plant and specification, the components communicate as described before via shared variables and label synchronization. Since $\mathcal{H}_\varphi$ defines the set of bad states, we use S_{bad} for the set of bad states both in $\mathcal{H}$ and in $\mathcal{H}_{BC} \times \mathcal{H}$.

To guarantee that the system satisfies the specification for unbounded time, the reachability tool searches for fixpoints outside of the bad states, that means it checks whether $reach_\infty^{\mathcal{H}\times\mathcal{H}_{BC}}(Init_{\mathcal{H}\times\mathcal{H}_{BC}}) \cap S_{bad} = \emptyset$. We refer to the set of states that has been proven safe for unbounded time by $S_{kr} \subseteq (Loc_\mathcal{H} \times Loc_{BC}, 2^{\mathbb{R}^{|Var_\mathcal{H}|+|Var_{BC}|}})$. The set S_{kr} needs to satisfy the Recoverable Region Invariance: it cannot contain any bad states and it has to be closed under the reachability relation.

Definition 2 (Recoverable Region Invariance). *A set S fulfills the Recoverable Region Invariance if $S \cap S_{bad} = \emptyset$ and $reach_\infty^{\mathcal{H}\times\mathcal{H}_{BC}}(S) \subseteq S$.*

If the reachability computation terminates with a fixpoint which does not nclude a bad state, Definition 2 is guaranteed.

ʼixpoint Detection. As fixpoint detection is a known problem for flowpipe-onstruction-based reachability analysis methods, several improvements were dded to increase the robustness of the approach and thus the chances to find a xpoint. Starting from a classical approach where a fixpoint is found whenever novel initial set after a discrete jump is fully contained inside a previously omputed state set we propose several improvements.

First, we augment the reachability analysis method with an interface to access xternal data sources for fixpoint detection. This lets us accumulate results over several runs and thus evolve S_{kr}. External data is stored efficiently in a tree-like structure similar to *octrees* [16] that subdivides the state space into cells for faster lookup. For each location, we create a tree whose nodes represent a hierarchical partition of the state space, i.e., nodes of layer $i+1$ are a partition of the nodes on layer i. Computed sets are stored in the leaves of this structure to enable faster lookup; if a cell is fully covered this information can be cached for faster results. A minor, but effective improvement for the aforementioned data structure is to only store initial sets instead of all sets of states that are computed to save memory and speed up the lookup when searching for fixpoints.

Often, novel initial sets S' are not fully contained in a single, previously computed initial set S_i but are still contained in the union of several of those sets $S' \subseteq \bigcup S_i$. We extend fixpoint detection to handle this case by iteratively checking for all S_i whether $S' = S'/S_i$ eventually becomes empty. Note that this check requires computing *set-difference*, which is hard for arbitrary state sets, e.g., convex polytopes, as the potentially non-convex result needs to be convexified afterwards. To overcome this, we fix our method to operate on boxes, which allows more efficient implementation of the set-difference operation.

Furthermore, in some cases, we could not find fixpoints due to *Zeno-behavior*, i.e., infinitely many discrete jumps in zero time. An example of such behavior can often be observed in *switched systems*, where the state space is partitioned into cells where the dynamics in each cell is described by a single location. For instance, having two neighboring locations ℓ, ℓ' connected by jumps with guards $x \geq 5$ and $x \leq 5$, for $x = 5$ the system can switch infinitely often between those locations without making any progress. To overcome this problem we have added detection for those Zeno-cycles that do not allow progress into our analysis method, such that these cycles get executed only once and thus can be declared a fixpoint. In contrast to the aforementioned approach to finding fixpoints, which operates on the computed state sets, this method analyzes cycles symbolically and does not cause over-approximation errors. Intuitively, for a path π leading to a reachable location ℓ, we iteratively compute sets of states that would possible allow Zeno behavior: initially, we consider the set of states satisfying the guard condition of the incoming transition to ℓ. Going back in the considered path, we alternatingly add constraints for invariant conditions and further incoming transitions along the locations and transition on path π while also adding transformations according to the reset functions on the transitions. This way, we can encode a symbolic expression representing the set of states, that enables Zeno-behavior. Checking containment of the actual state sets and the computed symbolic set allows to find Zeno-cycles of length up to the length of π.

Parallel Composition. Computation of S_{kr} is performed on the product of the plant-automaton, the specification-automaton, and the baseline controller automaton. To improve scalability, we feature an on-the-fly parallel composition that unrolls the product automaton during analysis as required. This improves execution speed and reduces the memory footprint.

Algorithm 1. Execution using a Simplex architecture.

1: $((\ell, \ell_B, \ell_\varphi), (X, U_B, X_\varphi)) \leftarrow Init_{H \times H_{BC}}$
2: **loop**
3: $\quad U_A \leftarrow AC(X)$
4: $\quad U_B \leftarrow BC(X)$
5: $\quad ((\ell', \ell'_B, \ell'_\varphi), (X', U'_B, X'_\varphi)) \leftarrow reach_{=\delta}^{H \times H_{BC}}(((\ell, \ell_B, \ell_\varphi), (X, U_A, X_\varphi)))$
6: $\quad suc \leftarrow ((\ell', \ell'_B, \ell'_\varphi), (X', U'_B, X'_\varphi)) \subseteq S_{kr}$
7: $\quad$ **if** $\neg suc$ **then**
8: $\quad\quad (suc, S_{kr}) \leftarrow \text{TRAIN}(S_{kr}, ((\ell', \ell'_B, \ell'_\varphi), (X', U'_B, X'_\varphi)))$
9: $\quad$ **end if**
10: $\quad$ **if** suc **then**
11: $\quad\quad X \leftarrow runPlant(U_A, \delta)$
12: $\quad$ **else**
13: $\quad\quad ((\ell', \ell'_B, \ell'_\varphi), (X', U_B, X'_\varphi)) \leftarrow reach_{=\delta}^{H \times H_{BC}}(((\ell, \ell_B, \ell_\varphi), (X, U_B, X_\varphi)))$
14: $\quad\quad X \leftarrow runPlant(U_B, \delta)$
15: $\quad$ **end if**
16: $\quad (\ell, \ell_B, \ell_\varphi, X_\varphi) \leftarrow (\ell', \ell'_B, \ell'_\varphi, X'_\varphi)$ ▷ Update for next loop iteration
17: **end loop**
18: **procedure** $\text{TRAIN}(S_{kr}, \sigma)$
19: $\quad S_{new} \leftarrow bloat(\sigma)$
20: $\quad$ **if** $reach_{\infty}^{H \times H_{BC}}(S_{new}, S_{kr}) \cap S_{bad} = \emptyset$ **then**
21: $\quad\quad$ **return** $\left(\top, S_{kr} \cup reach_{\infty}^{H \times H_{BC}}(S_{new}, S_{kr})\right)$
22: $\quad$ **else**
23: $\quad\quad$ **return** $(\bot, S_{kr})$
24: $\quad$ **end if**
25: **end procedure**

4.3 Simplex Execution with Proofs on Demand

Once a known recoverable region S_{kr} has been verified it can be used in a switching logic. The intuition is to analyze the predicted set of reachable states for the plant (and the specification) *ahead* for one control cycle and decide whether to use the advanced controller or the baseline controller. The decision is based on whether these results are compatible with the previously computed recoverable region S_{kr}, or if S_{kr} can be extended to allow for these states. In the following, we give a more technical description of this approach which is also shown in Algorithm 1.

The initial state is obtained from the *model* of the composition of plant and specification. In a loop, the method receives the suggested outputs U_A, U_B from both advanced and baseline controller (Lines 3 and 4) based on the current observable state X. Since advanced controller and baseline controller may be stateful, this step may also update their internal states based on (ℓ, X) during the computation of the controller output. In a next step, we use reachability analysis from the current state $((\ell, \ell_B, \ell_\varphi),\ (X, U_A, X_\varphi))$ to obtain all possible δ-reachable states $((\ell', \ell'_B, \ell'_\varphi), (X', U'_B, X'_\varphi))$ of the plant, the baseline controller, and the specification when using the output from the advanced controller (Line 5). The analysis is done for the length δ of one control cycle. Note that in this step, we analyze the composition of the plant, the specification, and the baseline controller using the output of the advanced controller to obtain all possible initial states for the next iteration. The idea is to be able to validate, whether after having invoked the advanced controller, the resulting configuration of the plant and the specification yields a configuration from which the baseline controller can ensure safety afterwards if required.

The system in its current state is *recoverable* when using the advanced controller, if the newly obtained states $((\ell', \ell'_B, \ell'_\varphi), (X', U'_B, X'_\varphi))$ are fully contained in S_{kr} (Line 6). The training function TRAIN can be invoked to attempt to extend S_{kr} (Line 8) if the new states are not yet included. If the new states are contained in S_{kr} (possibly after extending it), the plant will be run for one control cycle with the control input of the advanced controller (Line 11). Otherwise, the plant is executed using the baseline controller output (Line 14). In both cases, the state of the plant is observed and stored in X.

The procedure TRAIN(S_{kr}, σ) (Line 18) checks, if the new observation is safe for unbounded time. In this case, it returns the new recoverable region together with a Boolean flag $\top$. Otherwise, the procedure returns $\bot$ and the old recoverable region. The construction of the specification automaton ensures that a bad state can never be left, if any point between now and σ would unsafe so is σ. To produce results that generalize beyond a single state we *bloat* the state set by extending it in all dimensions to a configurable size. Unbounded time safety is checked for this enlarged set. This can be established either by proving that all trajectories reach S_{kr} or by finding a new fixpoint. Extending S_{kr} using TRAIN preserves recoverability (Definition 2).

Proposition 1. *Let S'_{kr} be the set computed by* TRAIN(S_{kr}, s) *then*

$$\begin{aligned}&(S_{kr} \cap S_{bad} = \emptyset \wedge reach_\infty^{H \times H_{BC}}(S_{kr}) \subseteq S_{kr})\\ \Rightarrow &(S'_{kr} \cap S_{bad} = \emptyset \wedge reach_\infty^{H \times H_{BC}}(S'_{kr}) \subseteq S'_{kr})\end{aligned}$$

The intersection of the states added by TRAIN and of S_{kr} with S_{bad} is empty. Thus, this also holds for S'_{kr}. There are two cases to show that the evolution of all states remains in S'_{kr}: First, the added states originate from a flowpipe that fully leads into S_{kr}. In this case, all added states will have a trajectory into S_{kr} when using the baseline controller, where they will stay by the assumption for S_{kr}. In the second case the new recoverable region that is added that has its own fixpoint, i.e., is safe for unbounded time. The used reachability method guarantees that every trajectory stays in this region which is a subset of S'_{kr}. Thus S'_{kr} satisfies both properties from Definition 2, i.e., a system controlled by Algorithm 1 satisfies φ, if its initial state is in S_{kr}.

The evolutionary nature of this approach allows to provide *proofs on demand* even during running time, provided the system environment is equipped with enough computational power to perform reachability analysis. Since this is in general not the case, the approach can be adapted to collect potential new initial sets S_{AC} and verify those offline or run verification asynchronously (online). In the later cases, since safety cannot directly be shown for S_{AC}, the system switches to using the baseline controller and results obtained offline or asynchronously can be integrated into future iterations.

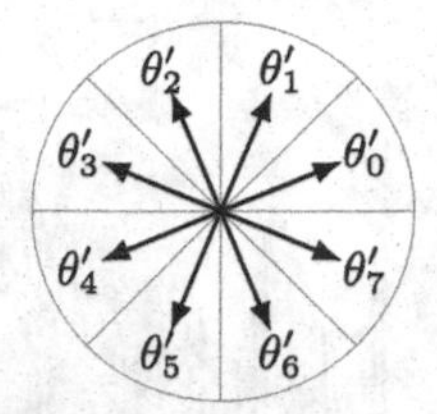

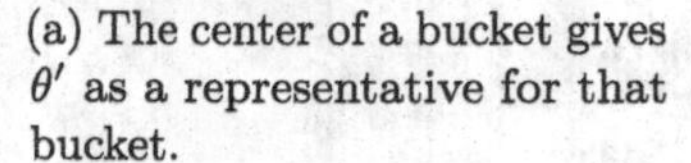
(a) The center of a bucket gives θ' as a representative for that bucket.

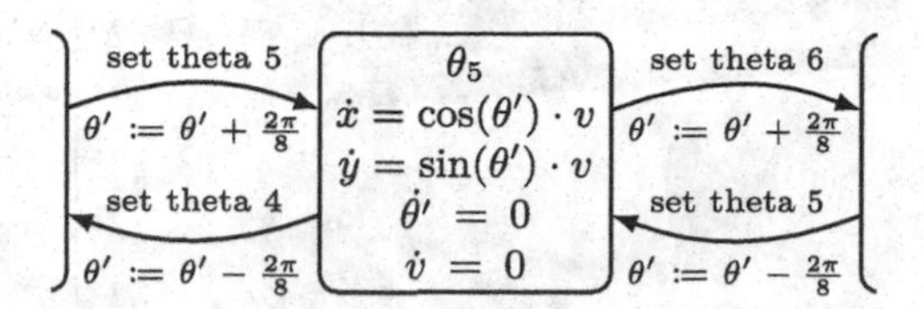

(b) Locations for different buckets are only connected by controller-actions (label synchronization), which are periodically enabled with cycle time δ.

Fig. 6. Modelling approach for the discretized car for 8 buckets.

5 Case Study: Autonomous Racing Car

We evaluate our approach to a controller for an autonomous racing car. The car is modeled as a point mass, observables are the position (x, y), the heading (θ), and its velocity (v). The car can be modeled by a hybrid automaton with a single location and non-linear dynamics. For simplification, we allow instantaneous changes of the velocity and do not model acceleration. To obtain a *linear* hybrid system, we discretize θ and replace it with a representative such that the transcendental terms become constants (see Fig. 6a); the number of buckets for this discretization is parameterized and induces multiple locations (one for each bucket for each discretized variable, see Fig. 6b).

The car is put on different circular racetracks where the safety specification is naturally given by the track boundaries. Each track is represented as three collections of convex polygonal shapes $P_{in}, P_{out}, P_{curbs} \subseteq \mathbb{R}^2$ which define the inner and outer boundary of the track (see yellow area in Fig. 7a), as well as the curbs on the border of the track. Whenever the car enters the curbs it must exit them within 2 time units. Formally, the specification is $\varphi = \mathbf{G}((x, y) \notin P_{in}) \wedge \mathbf{G}((x, y) \notin P_{out}) \wedge \mathbf{G}(((x, y) \in P_{curbs}) \rightarrow \mathbf{F}_{[0,2]}((x, y) \notin P_{curbs}))$.

Baseline Controller. To model the baseline controller each track is subdivided into an ordered sequence of straight segments. The control inputs of the baseline controller attempt to drive the car to the center region of the current segment where the car stops. To model this behavior, each segment is subdivided into several zones with different dynamics, depending on the relative position of the zone to the center of the segment.

Advanced Controller. The advanced controller implements a pure pursuit ontroller [3] that is equipped with a set of waypoints along the track. Waypoints re either given by points in the middle of the track on the boundary between wo segments or in a more advanced setup obtained by a raceline optimizer ool [9].

Results and Observations. For evaluation, we consider several tracks that are ther simple toy examples such as square- or L-shaped tracks or linearizations actual F1 racetracks.

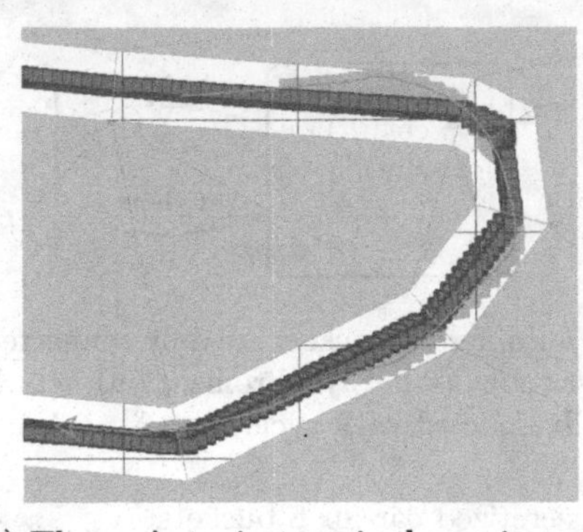

(a) The car's trajectory is shown in green. Leaving S_{kr} (dark blue) triggers extension thereof (light blue).

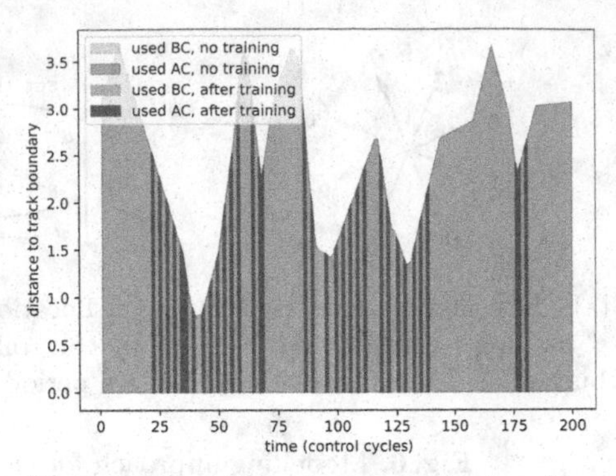

(b) Distance to the closest track boundary over time. Controller usages are color-coded for each control cycle.

Fig. 7. Execution of the race car with online verification. (Color figure online)

We outline some of the results that we obtained during evaluation here, we provide a full set of images and videos online[1].

The designed baseline controller is relatively conservative, it prioritizes safety over progress as it steers the car toward the center line of the track and then stops. This behavior enables fast computation of unbounded safety results but as a drawback does not allow for large extensions of S_{kr} during a single training run. To show the influence of proofs on demand, we synthesized S_{kr} for the center area of a whole track a priori. Depending on the selection of waypoints for the advanced controller, naturally, the usefulness of the initial S_{kr} varies. Waypoints in the middle tend to induce fewer extensions of S_{kr} (or, fewer invocations of the baseline controller in case proofs on demand are disabled) than the set of waypoints generated by the raceline optimizer, which selects points allowing a path with less curvature. A visualization of the optimized trajectory, S_{kr}, as well as its extension can be seen in Fig. 7a.

We recorded the number of invocations of adaptive proofs over time (see e.g. Figure 7b), the results are plotted in Fig. 8 for the two sets of waypoints. We can observe, that initially many proofs are required to saturate S_{kr} in the first lap. Later laps require fewer adaptations, the number of proofs stabilizes at around 40 proofs per lap. This results from the controller not ending up perfectly in its starting position from the lap before, such that evolving S_{kr} is necessary to use the advanced controller as often as possible. The initial position used to generate proofs on demand is bloated to allow larger extensions of S_{kr} at a time. Our experiments with different bloatings (Table 1) show two opposing effects with increasing bloating: (1) fewer requests for extensions due to the larger extension, and (2) more invocations of the baseline controller as fewer extensions are successful due to the attempt of more aggressive expansion. This observation is in line with our expectations, as attempts for more aggressive extension of S_{kr} due to larger bloating may lead, if successful, to less requests for further

[1] https://github.com/modass/simplex-architectures/wiki/Experimental-results.

Table 1. Bloatings used for proofs on demand affect the success of this approach.

bloating	BC invocations	extensions
0.25	123	2945
0.5	142	1329
0.75	189	1133
1.0	268	803

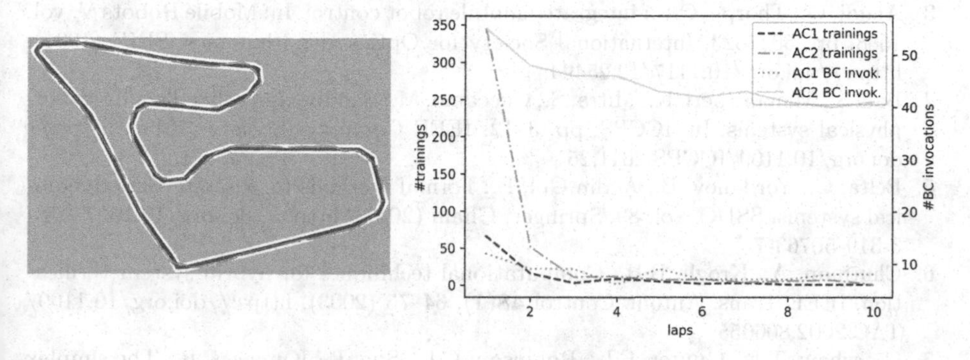

Fig. 8. Successful extensions of S_{kr} (left: blue) per lap for the F1 track. Starting from a moderately large known recoverable region for two sets of waypoints (standard = AC1, optimized = AC2). (Color figure online)

extensions. Otherwise, if those attempts are not successful, the consequence are more invocations of the baseline controller. A user can choose an appropriate trade off between verification time and how conservative the safety check is.

On a standard desktop computer, the computation of the initial recoverable region takes about 20 min, performing a proof-on-demand to extend the recoverable region takes less than a second, testing whether to switch without extending takes about 10 ms.

6 Conclusion

We have presented a method to incorporate proofs on demand in a fully automated Simplex architecture toolchain to ensure controllers obey a given temporal specification. Our method operates incrementally, fitting the recoverable region of the baseline controller to the behavior of the running advanced controller. Since our Simplex architecture adapts to the advanced controller, it allows performance increases by avoiding unnecessary switches to the baseline controller and invokes fewer verification queries at later stages of the execution phase.

One direction for future work is to use the robustness values of the STL specification to fine-tune the switching mechanism from the baseline controller to the advanced controller. Furthermore, we want to combine our Simplex architecture with reinforcement learning such that the architecture guides the learning phase via reward shaping and, at the same time, ensures correctness during training.

References

1. Alshiekh, M., Bloem, R., Ehlers, R., Könighofer, B., Niekum, S., Topcu, U.: Safe reinforcement learning via shielding. In: AAAI, pp. 2669–2678. AAAI Press (2018)
2. Alur, R., Courcoubetis, C., Henzinger, T.A., Ho, P.-H.: Hybrid automata: an algorithmic approach to the specification and verification of hybrid systems. In: Grossman, R.L., Nerode, A., Ravn, A.P., Rischel, H. (eds.) HS 1991-1992. LNCS, vol. 736, pp. 209–229. Springer, Heidelberg (1993). https://doi.org/10.1007/3-540-57318-6_30
3. Amidi, O., Thorpe, C.E.: Integrated mobile robot control. In: Mobile Robots V. vol. 1388, pp. 504–523. International Society for Optics and Photonics, SPIE (1991). https://doi.org/10.1117/12.25494
4. Bak, S., Manamcheri, K., Mitra, S., Caccamo, M.: Sandboxing controllers for cyber-physical systems. In: ICCPS, pp. 3–12. IEEE Computer Society (2011). https://doi.org/10.1109/ICCPS.2011.25
5. Belta, C., Yordanov, B., Aydin Gol, E.: Formal methods for discrete-time dynamical systems. SSDC, vol. 89. Springer, Cham (2017). https://doi.org/10.1007/978-3-319-50763-7
6. Chutinan, A., Krogh, B.H.: Computational techniques for hybrid system verification. IEEE Trans. Autom. Control **48**(1), 64–75 (2003). https://doi.org/10.1109/TAC.2002.806655
7. Crenshaw, T.L., Gunter, E.L., Robinson, C.L., Sha, L., Kumar, P.R.: The simplex reference model: limiting fault-propagation due to unreliable components in cyber-physical system architectures. In: RTSS, pp. 400–412. IEEE Computer Society (2007). https://doi.org/10.1109/RTSS.2007.34
8. Frehse, G., et al.: A toolchain for verifying safety properties of hybrid automata via pattern templates. In: ACC, pp. 2384–2391. IEEE (2018). https://doi.org/10.23919/ACC.2018.8431324
9. Heilmeier, A., Wischnewski, A., Hermansdorfer, L., Betz, J., Lienkamp, M., Lohmann, B.: Minimum curvature trajectory planning and control for an autonomous race car. Vehicle Syst. Dynam. **58**(10), 1497–1527 (2020). https://doi.org/10.1080/00423114.2019.1631455
10. Henzinger, T.A.: The theory of hybrid automata. In: Inan, M.K., Kurshan, R.P. (eds.) Verification of Digital and Hybrid Systems. NATO ASI Series, vol. 170, pp. 265–292. Springer, Heidelberg (2000). https://doi.org/10.1007/978-3-642-59615-5_13
11. Henzinger, T.A., Kopke, P.W., Puri, A., Varaiya, P.: What's decidable about hybrid automata? J. Comput. Syst. Sci. **57**(1), 94–124 (1998). https://doi.org/10.1006/jcss.1998.1581
12. Ionescu, T.B.: Adaptive simplex architecture for safe, real-time robot path planning. Sensors **21**(8), s21082589 (2021). https://doi.org/10.3390/s21082589
13. Johnson, T.T., Bak, S., Caccamo, M., Sha, L.: Real-time reachability for verified simplex design. ACM Trans. Embed. Comput. Syst. **15**(2), 1–27 (2016). https://doi.org/10.1145/2723871
14. Maler, O., Nickovic, D.: Monitoring temporal properties of continuous signals. In: Lakhnech, Y., Yovine, S. (eds.) FORMATS/FTRTFT -2004. LNCS, vol. 3253, pp. 152–166. Springer, Heidelberg (2004). https://doi.org/10.1007/978-3-540-30206-3_12
15. Marta, D., Pek, C., Melsión, G.I., Tumova, J., Leite, I.: Human-feedback shield synthesis for perceived safety in deep reinforcement learning. IEEE Robotics Autom. Lett. **7**(1), 406–413 (2022). https://doi.org/10.1109/LRA.2021.3128237

16. Meagher, D.: Geometric modeling using octree encoding. Comput. Graphics Image Process. **19**(2), 129–147 (1982). https://doi.org/10.1016/0146-664X(82)90104-6
17. Mehmood, U., Stoller, S.D., Grosu, R., Roy, S., Damare, A., Smolka, S.A.: A distributed simplex architecture for multi-agent systems. In: Qin, S., Woodcock, J., Zhang, W. (eds.) SETTA 2021. LNCS, vol. 13071, pp. 239–257. Springer, Cham (2021). https://doi.org/10.1007/978-3-030-91265-9_13
18. Mehmood, U., D. Stoller, S., Grosu, R., A. Smolka, S.: Collision-free 3D flocking using the distributed simplex architecture. In: Bartocci, E., Falcone, Y., Leucker, M. (eds.) Formal Methods in Outer Space. LNCS, vol. 13065, pp. 147–156. Springer, Cham (2021). https://doi.org/10.1007/978-3-030-87348-6_9
19. Phan, D.T., Grosu, R., Jansen, N., Paoletti, N., Smolka, S.A., Stoller, S.D.: Neural simplex architecture. In: Lee, R., Jha, S., Mavridou, A., Giannakopoulou, D. (eds.) NFM 2020. LNCS, vol. 12229, pp. 97–114. Springer, Cham (2020). https://doi.org/10.1007/978-3-030-55754-6_6
20. Prajna, S., Jadbabaie, A.: Safety verification of hybrid systems using barrier certificates. In: Alur, R., Pappas, G.J. (eds.) HSCC 2004. LNCS, vol. 2993, pp. 477–492. Springer, Heidelberg (2004). https://doi.org/10.1007/978-3-540-24743-2_32
21. Romdlony, M.Z., Jayawardhana, B.: Stabilization with guaranteed safety using control Lyapunov-barrier function. Automatica **66**, 39–47 (2016). https://doi.org/10.1016/j.automatica.2015.12.011
22. Schupp, S.: State set representations and their usage in the reachability analysis of hybrid systems, Ph. D. thesis, RWTH Aachen University, Aachen (2019). https://doi.org/10.18154/RWTH-2019-08875
23. Schupp, S., et al.: Current challenges in the verification of hybrid systems. In: Berger, C., Mousavi, M.R. (eds.) CyPhy 2015. LNCS, vol. 9361, pp. 8–24. Springer, Cham (2015). https://doi.org/10.1007/978-3-319-25141-7_2
24. Schupp, S., Ábrahám, E., Makhlouf, I.B., Kowalewski, S.: HyPro: A C++ library of state set representations for hybrid systems reachability analysis. In: Barrett, C., Davies, M., Kahsai, T. (eds.) NFM 2017. LNCS, vol. 10227, pp. 288–294. Springer, Cham (2017). https://doi.org/10.1007/978-3-319-57288-8_20
25. Seto, D., Krogh, B., Sha, L., Chutinan, A.: The simplex architecture for safe online control system upgrades. In: ACC, pp. 3504–3508. IEEE (1998). https://doi.org/10.1109/ACC.1998.703255
26. Sha, L.: Using simplicity to control complexity. IEEE Softw. **4**, 20–28 (2001). https://doi.org/10.1109/MS.2001.936213
27. Shivakumar, S., Torfah, H., Desai, A., Seshia, S.A.: SOTER on ROS: a run-time assurance framework on the robot operating system. In: Deshmukh, J., Ničković, D. (eds.) RV 2020. LNCS, vol. 12399, pp. 184–194. Springer, Cham (2020). https://doi.org/10.1007/978-3-030-60508-7_10

28. Simão, T.D., Jansen, N., Spaan, M.T.J.: Alwayssafe: reinforcement learning without safety constraint violations during training. In: Dignum, F., Lomuscio, A., Endriss, U., Nowé, A. (eds.) AAMAS 2021: 20th International Conference on Autonomous Agents and Multiagent Systems, Virtual Event, United Kingdom, 3–7 May 2021, pp. 1226–1235. ACM (2021). https://doi.org/10.5555/3463952.3464094
29. Yang, J., Islam, M.A., Murthy, A., Smolka, S.A., Stoller, S.D.: A simplex architecture for hybrid systems using barrier certificates. In: Tonetta, S., Schoitsch, E., Bitsch, F. (eds.) SAFECOMP 2017. LNCS, vol. 10488, pp. 117–131. Springer, Cham (2017). https://doi.org/10.1007/978-3-319-66266-4_8

Explicit-State Model Checking

Elimination of Detached Regions in Dependency Graph Verification

Peter Gjøl Jensen, Kim Guldstrand Larsen, Jiří Srba, and Nikolaj Jensen Ulrik(✉)

Department of Computer Science, Aalborg University, Aalborg, Denmark
{pgj,kgl,srba,njul}@cs.aau.dk

Abstract. The formalism of dependency graphs by Liu and Smolka is a well-established method for on-the-fly computation of fixed points over Boolean domains with numerous applications in e.g. model checking, game synthesis, bisimulation checking and others. The original Liu and Smolka on-the-fly algorithm runs in linear time, and several extensions and improvements to this algorithm have recently been studied, including the notion of negation edges, certain-zero early termination as well as extensions towards abstract dependency graphs. We present a novel improvement for computing the least fixed-point assignment on dependency graphs, with the goal of avoiding the exploration of detached subgraphs that cannot contribute to the assignment value of the root node. We also experimentally evaluate different ways of resolving nondeterminism in the algorithm and execute experiments on CTL formulae from the annual Petri net model checking contest as well as on synthesis problems for Petri games. We find out that our algorithm significantly improves the state-of-the-art.

1 Introduction

Within the fields of formal verification and model checking, the efficient computation of fixed points is of great importance for solving many problems, such as
bisimulation checking [26] or model checking Computation Tree Logic (CTL) [7]
or the modal μ-calculus [22]. Unfortunately, the naive approaches to computing
such fixed points are prone to state space explosion as they require the entire
state space to be available before verification can be done, which is often not a
easible option.

The formalism of Dependency Graphs (DGs), developed by Liu and
molka [24], is a general formalism that encodes dependencies among the differ-
nt nodes in the graph and allows for efficient, on-the-fly fixed-point computa-
ons. A *dependency graph* is a directed graph with hyperedges, i.e. edges that can
ave multiple targets. Figure 1b shows an example of a dependency graph that
ncodes the problem of checking whether the Kripke structure shown in Fig. 1a
atisfies the CTL formula $\mathsf{E}(a \vee b)\, \mathsf{U}\, c$ using the encoding of [10]. Hyperedges are
rawn as branching from a mid-point, for example there is a hyperedge from v_0
ith v_2 and v_3 as targets. An *assignment* of the dependency graph is a function

Caltais and C. Schilling (Eds.): SPIN 2023, LNCS 13872, pp. 163–179, 2023.
ps://doi.org/10.1007/978-3-031-32157-3_9

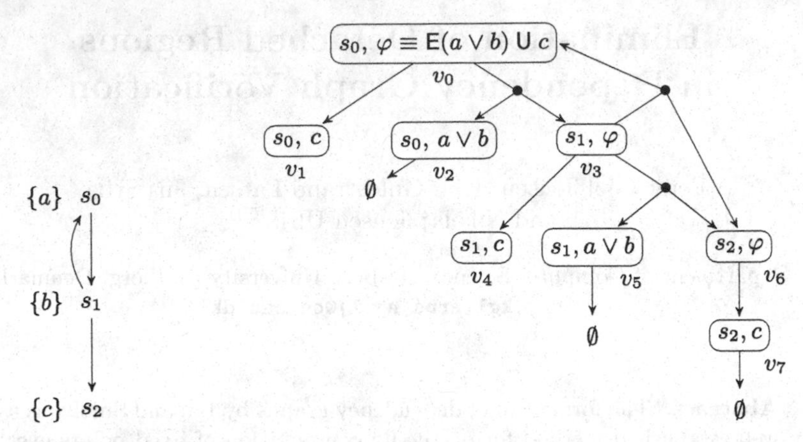

(a) Kripke Structure (b) DG encoding of model checking $s_0 \models \mathsf{E}(a \vee b) \mathsf{U}\, c$

Iteration	$A(v_0)$	$A(v_1)$	$A(v_2)$	$A(v_3)$	$A(v_4)$	$A(v_5)$	$A(v_6)$	$A(v_7)$
0	0	0	0	0	0	0	0	0
1	0	0	1	0	0	1	0	1
2	0	0	1	0	0	1	1	1
3	0	0	1	1	0	1	1	1
4	1	0	1	1	0	1	1	1
5	1	0	1	1	0	1	1	1

(c) Fixed-point computation on the DG

Fig. 1. Example of dependency graph encoding a CTL model checking problem

that assigns to each node in the graph the value 0 (false) or 1 (true). The goal of dependency graph verification is to compute the minimum fixed-point assignment A (given by Tarski's fixed point theorem [27]). This computation is shown in Fig. 1c by starting from the assignment where all nodes have the value 0 and iteratively performing the following saturation procedure: if there is a hyperedge from a node v such that all the targets of the hyperedge already have the value 1, then v also gets the value 1. A hyperedge with the empty set of targets is interpreted as a vacuous truth and propagates the value 1 to its source, whereas if a node does not have any outgoing hyperedges, its value remains 0. We can see that the saturation procedure stabilizes after the fourth iteration and reaches the minimum fixed point where $A(v_0) = 1$, meaning that $s_0 \models \mathsf{E}(a \vee b) \mathsf{U}\, c$.

The success of dependency graphs is due to an efficient, on-the-fly and local algorithm for computing the least fixed-point assignment [24], meaning that the full state space is not necessarily needed before the fixed-point computation can start. More recently, a distributed version of the algorithm was developed [8], and the formalism of dependency graphs was extended to support negation edges [9]. Dependency graphs have shown great success in bisimulation checking [2] and model checking of recursive Hennessy-Milner logic [2] and CTL [10], and variations of dependency graphs have been applied to strategy synthesis of Timed Games [6] and model checking of weighted CTL [18], probabilistic CTL [25] and the modal μ-calculus [23]. For an overview of different applications of dependency graphs to model checking, consult [13].

In this paper, we present a novel improvement to the dependency graph verification algorithm. Our contribution is two-fold. First, we extend the dependency graph algorithm with an optimization that allows us to prune detached regions of the dependency graph (parts of the graph that were scheduled for exploration but which during the fixed-point computation became irrelevant for the fixed-point assignment of the root node) and hence to improve the running time of the computation. We experimentally demonstrate that this improvement significantly helps in solving CTL model checking and games synthesis problems. Second, we investigate and discuss in detail the impact of different choices of implementation of the general nondeterministic algorithm, in particular wrt. the search order used by the algorithm, and provide an extensive experimential evaluation.

Related Work. Dependency graphs found great success through Liu and Smolka's on-the-fly, linear-time algorithm for computing fixed-points [24]. This algorithm includes early termination when a value of 1 is computed for the designated root node of the dependency graph. Early termination was later extended with the certain-zero optimization [10], which also allows propagation of the value 0 in cases where it provably cannot be improved anymore. Other improvements and extensions include negation edges [10], distributed verification [8] of dependency graphs and most recently the development of abstract dependency graphs [15]. However, most of these additions do not address the issue that the algorithm has a risk of spending time exploring detached regions of the dependency graphs, i.e. regions where improving the fixed-point assignment will not lead to improvement of the root node value. The algorithm of [15] attempts to address the detached regions issue, however, in a manner which proved inefficient in practice (requiring a potentially expensive recursive propagation of information about detached regions). In contrast, our method of addressing the problem captures a high number of detached regions but with a minimal overhead.

Dependency graphs are used by the verification engines of several tools. The tool CAAL [2] uses dependency graphs for bisimulation checking and model checking of recursive Hennessy-Milner logic on CCS [26] processes. UPPAAL TiGa [3] supports controller synthesis for games on Timed Automata [1], and the tool TAPAAL [11] supports strategy synthesis for Petri net games as well as CTL model checking on Petri nets, all using dependency graphs. Additionally, dependency graphs have been applied to probabilistic [25] and weighted [18] domains. The formalism of Boolean Equations Systems, which are very similar to dependency graphs, are used for various verification tasks within the μCRL2 [5] and CADP [16] tools. We believe that our technique for detached regions detection can contribute to improved verification performance of such tools.

In [14], it was observed that the basic principles behind the dependency graph algorithms used in different application areas are not very different, except that they generalize the assignments from Boolean values to more complex domains, for example to weighted domains. This prompted the theory of abstract dependency graphs, which encompasse all the aforementioned application areas. In [15] this formalism is extended to include nonmonotonicity, which generalises the notion of negation edge which was developed for CTL model checking in [10]. Our work focuses on the traditional dependency graph formalism with assignments over Boolean domain, leaving the extension towards more general domains for future work.

2 Dependency Graphs

We shall now introduce the formalism of dependency graphs together with Boolean valued assignments to its nodes and define the minimum fixed-point assignment.

Definition 1 (Dependency graph). *A* DG *is a pair $G = (V, E)$ where V is a finite, nonempty set of* configurations *(nodes) and $E \subseteq V \times 2^V$ is a set of* hyperedges.

For a hyperedge $e = (v, T) \in E$, we call v the *source* of e and $u \in T$ the *targets* of e. We let $\text{sucs}(v)$ denote the set of hyperedges with v as the source configuration, i.e. $\text{sucs}(v) = \{(v, T) \mid (v, T) \in E\}$.

Each configuration of a DG is associated with a Boolean function defined by a disjunction over each hyperedge, and each hyperedge is a conjunction over the set of target configurations. The semantics of a DG is given by Boolean-valued assignments such that the Boolean function on each configuration is satisfied.

Definition 2 (Assignment). *Let $G = (V, E)$ be a DG. An* assignment *on G is a function $A : V \to \{0, 1\}$, and the set of all assignments on G is denoted $\mathcal{A}^G$. For $A, A' \in \mathcal{A}^G$, let $A \preceq A'$ iff for all $v \in V$ we have $A(v) \leq A'(v)$, and let $A_\perp$ denote the assignment such that $A_\perp(v) = 0$ for all $v \in V$. An assignment A on G is a* solution *(fixed-point assignment) iff for all $v \in V$ we have*

$$A(v) = \bigvee_{(v,T)\in E} \bigwedge_{u\in T} A(u)$$

where we define the empty conjunction to be 1 *(true) and the empty disjunction to be* 0 *(false).*

Note that for any DG G, $(\mathcal{A}^G, \preceq)$ is a complete lattice with least element $A_\perp$. We can express solutions as fixed points of the function $F : \mathcal{A}^G \to \mathcal{A}^G$ defined by

$$F(A)(v) = A(v) \vee \bigvee_{(v,T)\in E} \bigwedge_{u\in T} A(u) \ . \qquad (1)$$

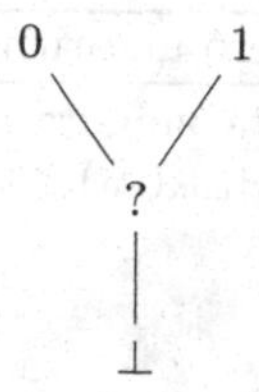

Fig. 2. Illustration of the certain-zero partial order.

Clearly F is monotonic, so by Tarski's fixed-point theorem [27] the function F has a minimum fixed point, and furthermore there exists an $i \geq 0$ such that $F^i(A_\perp) = F^{i+1}(A_\perp)$, in which case $F^i(A_\perp)$ is a minimal solution on G. We write $A_{\min}$ to refer to the minimum fixed-point solution $F^i(A_\perp)$. An example of the iterative computation of the minimum fixed point (also referred to as the global algorithm) is depicted in Fig. 1c.

Remark 1. For sake of simplicity, we omit methods of dealing with negation in verified properties. The paper [10] extends DGs with negation edges and shows a correct encoding of CTL properties, assuming that negation edges are explored exactly when all configurations below have been explored, and only then. Our experimental evaluation includes a benchmark of CTL queries, so our implementation supports negation edges. As also shown in [15], this works near trivially if the DG is explored depth-first. We refer to [10] and [15] for a proper treatment of negation.

3 Local Algorithm with Detached Regions Detection

For a DG $G = (V, E)$, our goal is to compute $A_{\min}(v_0)$ for a distinguished configuration $v_0 \in V$. The original algorithm for doing this efficiently was published by Liu and Smolka [24] and later on extended with the notion of certain-zero [10], demonstrating significant improvement in verification times. The Liu and Smolka algorithm explores the dependency graph in a forward manner, starting from the root v_0, and backpropagates the value 1 along the hyperedges whenever possible. The certain-zero extension allows the algorithm to also backpropagate the value 0 when it can prove that the value of a configuration cannot be improved to 1 anymore. In effect, the certain-zero improvement extends the Boolean domain to the simple partial order of assignment values depicted in Fig. 2. Here, the value $\perp$ means that a configuration has not been discovered yet, ? means that a discovered configuration waits to be processed but its value is not determined yet, and the values 1 and 0 are final configuration values in the minimum fixed-point assignment. Our new improvement to the algorithm is an additional check whether a configuration can be ignored in the situations where its backpropagation only affects configurations that are already fully resolved.

Algorithm 1 shows the procedure in pseudocode. If all lines highlighted in gray are removed, then the algorithm corresponds to the original algorithm by

Algorithm 1. Improved certain-zero algorithm

```
    Input: DG G = (V, E) and a root configuration v0 ∈ V.
    Output: Least fixed-point value A_min(v0).
 1: W ← sucs(v0)
 2: for all v ∈ V do
 3:     D(v) ← ∅
 4:     A(v) ← ⊥
 5: A(v0) ← ?
 6: while W ≠ ∅ do
 7:     e ← (v, T) ∈ W
 8:     W ← W \ {e}
 9:     if v ≠ v0 and (A(v) ∈ {0,1} or ∀(w, T') ∈ D(v) . A(w) ∈ {0,1}) then
10:         if A(v) ∉ {0,1} then A(v) ← ⊥
11:         goto line 6
12:     if ∀u ∈ T . A(u) = 1 then
13:         if v = v0 then return 1
14:         if A(v) ≠ 1 then
15:             A(v) ← 1; W ← W ∪ D(v)
16:     elseif ∃u ∈ T . A(u) = 0
17:         if A(v) ≠ 0 then
18:             E ← E \ {e}
19:             if sucs(v) = ∅ then
20:                 if v = v0 then return 0
21:                 A(v) ← 0; W ← W ∪ D(v)
22:     else
23:         pick u ∈ T such that A(u) ∉ {0,1}
24:         D(u) ← D(u) ∪ {e}
25:         if A(u) = ⊥ then
26:             A(u) ← ?
27:             if sucs(u) = ∅ then
28:                 if u = v0 then return 0
29:                 A(u) ← 0; W ← W ∪ D(u)
30:             else
31:                 W ← W ∪ {(u, T') | (u, T') ∈ E}
32: return 0                          ▷ v0 did not receive value, so must be 0.
```

Liu and Smolka. If the light gray lines (lines 16–21 and 27–29) are included, we obtain a single-core version of the certain zero optimization suggested in [10]. The dark gray lines 9–11 are our additional optimization to the algorithm.

The algorithm maintains a waiting list of hyperedges to be explored as well as for each $v \in V$ a set $D(v)$ of parent hyperedges that are dependent on the value of v. In the initialization step, all outgoing edges $(v_0, T) \in E$ from v_0 are pushed to the waiting-list, and $D(v)$ is initialised to the empty set for all $v \in V$. During each iteration of the **while**-loop at line 6, an edge $e = (v, T)$ is picked from the waiting list. If all the target configurations $u \in T$ have assignment 1 on line 12 then the value is assigned to $A(v)$ and the set $D(v)$ of hyperedges dependent on v are inserted into the waiting list. If some target configuration $u \in T$ has value 0 (certain-zero) on line 16 then the edge is deleted from the DG, and if now v has no outgoing edge, a value of 0 is assigned to v and all dependents in $D(v)$ are added to the waiting-list. If the configuration that was just assigned a final value is v_0 then we can terminate early, on lines 13 and 20. As observed in [13], the checks on lines 14 and 17 are important to ensure termination of the algorithm. If neither of the above cases were true, there must be some successor node $u \in T$ such that $A(u) \in \{?, \bot\}$, among which we select one on line 23. We update $D(u)$ to also contain e, and if $A(u) = \bot$, i.e. it is not discovered yet, we set $A(u) \leftarrow ?$ and push all outgoing edges $(u, T') \in E$ to the waiting list.

The lines 9–11 are our proposed improvement. The motivation is that we want to avoid spending time exploring detached regions of the dependency graph, i.e. configurations that cannot possibly help conclude the value of v_0 at the current point in execution. The test we implement here checks whether all edges $e = (u, T) \in D(v)$ have a final value 1 or 0 assigned to their source, as this means the value of v will not contribute to any further knowledge about the fixed-point assignment on G. In this case, we simply skip the edge, remove it from the waiting list. Since there may be previously unknown edges that will need the value $A(v)$ in a later iteration, we set $A(v) = \bot$ to indicate that it should be expanded again when found in the future during the forward search.

Remark 2. We can formalize the notion of detached regions as follows. Let *Dep* : $V \times V$ be a relation such that for $v, u \in V$, $Dep(v, u)$ holds if and only if there is $(u, T) \in D(v)$ such that $A(u) \notin \{0, 1\}$. We say that $v \in V$ is *detached* if $(v, v_0) \notin Dep^*$, where Dep^* is the transitive closure of *Dep*. Conversely, $v \in V$ is not detached if $(v, v_0) \in Dep^*$, indicating that there is some chain of dependencies from v to v_0 that does not contain any configurations that are assigned 0 or 1.

The test on lines 9–11 does *not* detect all detached regions. One approach to detect all detached regions can be obtained with a slight modification to Algorithm 1 of [15], calling UPDATEDEPENDENTS on line 11 of that algorithm. We implemented this in our tool, measuring significantly worse results than even the Certain-Zero algorithm, as we will detail in Sect. 5.

Example 1. Figure 3 shows different means of computing the least fixed-point value $A_{\min}(v_0)$ of the DG depicted in Fig. 3a. Figure 3b shows the computation of the minimum fixed-point assignment using the global algorithm given by 1. Each row corresponds to one iteration of the algorithm, terminating on iteration since $F^4(A_\bot) = F^3(A_\bot)$. We notice that in the global algorithm we iterate ver all nodes in the dependency graph.

Figure 3c shows the computation using the local algorithm with the certain-zero extension and Fig. 3d shows the computation using our improved version f the local algorithm, each with the same search order (the choice being the ght-most available configuration, and nodes v with $A(v) = ?$ have priority).

For iterations 1–4, the certain-zero algorithm simply searches through the aph in a forward fashion. Once e_3 is checked in iteration 3, we set $A(a) = 1$

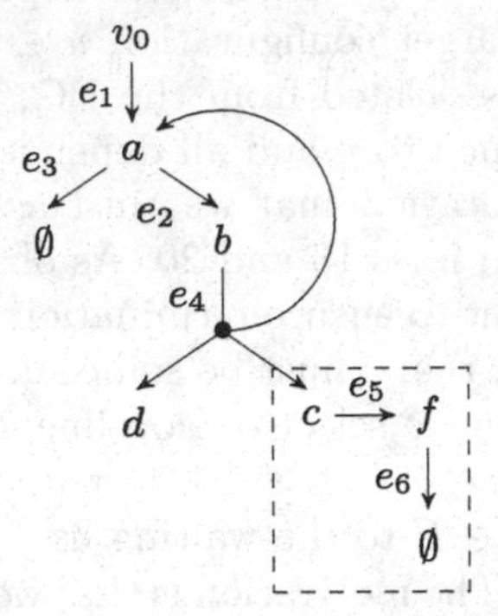

(a) Dependency graph (the dashed region eventually becomes detached)

Iteration	$A(v_0)$	$A(a)$	$A(b)$	$A(c)$	$A(d)$	$A(f)$
0	0	0	0	0	0	0
1	0	1	0	0	0	1
2	1	1	0	1	0	1
3	1	1	0	1	0	1
4	1	1	0	1	0	1

(b) Execution of the global algorithm

Iteration	W	$e \in (v, T)$	$u \in T$	$A(v_0)$	$A(a)$	$A(b)$	$A(c)$	$A(d)$	$A(f)$
0	$\{e_1\}$			?	$\bot$	$\bot$	$\bot$	$\bot$	$\bot$
1	$\{e_1\}$	e_1	a	?	?	$\bot$	$\bot$	$\bot$	$\bot$
2	$\{e_2, e_3\}$	e_2	b	?	?	?	$\bot$	$\bot$	$\bot$
3	$\{e_3, e_4\}$	e_4	a	?	?	?	$\bot$	$\bot$	$\bot$
4	$\{e_3\}$	e_3	—	?	1	?	$\bot$	$\bot$	$\bot$
5	$\{e_4, e_1\}$	e_4	c	?	1	?	?	$\bot$	$\bot$
6	$\{e_1, e_5\}$	e_5	f	?	1	?	?	$\bot$	?
7	$\{e_1, e_6\}$	e_6	—	?	1	?	?	$\bot$	1
8	$\{e_1, e_5\}$	e_5	—	?	1	?	1	$\bot$	1
9	$\{e_1, e_4\}$	e_4	d	?	1	?	1	0	1
10	$\{e_1, e_4\}$	e_4	—	?	1	0	1	0	1
11	$\{e_1\}$	e_1	—	1	1	?	1	0	1

(c) Iterations of the local certain-zero algorithm

Iteration	W	$e \in (v, T)$	$u \in T$	$A(v_0)$	$A(a)$	$A(b)$	$A(c)$	$A(d)$	$A(f)$
0	$\{e_1\}$			?	$\bot$	$\bot$	$\bot$	$\bot$	$\bot$
1	$\{e_1\}$	e_1	a	?	?	$\bot$	$\bot$	$\bot$	$\bot$
2	$\{e_2, e_3\}$	e_2	b	?	?	?	$\bot$	$\bot$	$\bot$
3	$\{e_3, e_4\}$	e_4	a	?	?	?	$\bot$	$\bot$	$\bot$
4	$\{e_3\}$	e_3	—	?	1	?	$\bot$	$\bot$	$\bot$
5	$\{e_4, e_1\}$	e_4	—	?	1	?	$\bot$	$\bot$	$\bot$
6	$\{e_1\}$	e_1	—	1	1	?	$\bot$	$\bot$	$\bot$

(d) Iterations of our improved local algorithm

Fig. 3. Dependency graph and fixed-point computations. The search order is depth first, choosing the right-most hyperedge and rightmost configuration in each step.

and re-add edges e_1 and e_4 to the waiting list in hopes that the value can be backpropagated further. The certain-zero algorithm picks the configuration c to be expanded. This enters the region consisting of the configurations c and f, which is detached since all paths back to v_0 contain some configuration u such that $A(u) \in \{0,1\}$. From c the search continues through edges e_5 and e_6, which eventually backpropagates the value 1 to $A(c)$. Finally on iteration 9, the configuration d is expanded and receives a value of certain-zero since it has no outgoing hyperedges, and on iteration 10 this is backpropagated to b.

On the other hand, our improved algorithm detects on iteration 5 that all configurations depending on b have been assigned, so there is no need to evaluate the edge e_4. Thus it avoids exploring the detached region, saving a total of 5 iterations over the certain-zero algorithm. We also notice that while the global algorithm takes fewer iterations to complete than the local algorithms, in each iteration the assignment of each configuration is checked, which is an expensive operation if the DG is large.

4 Correctness of the Algorithm

We now proceed to show correctness of the full updated algorithm, i.e. the algorithm including all lines, starting with some technical lemmas. The proofs are adapted from the certain-zero algorithm [10], with several differences that originate from the improvements in our algorithm, in particular from the possibility to assign the value $\bot$ to nodes that already had the value ? before.

Lemma 1. *For each $v \in V$, $A(v)$ is assigned a value $x \in \{0,1\}$ at most once.*

Proof. First we observe that for any configuration $v \in V$, whenever $A(v) \in \{0,1\}$, neither ? nor $\bot$ are ever assigned to it, since the value ? is only assigned on line 26, which can only be reached if $A(v) = \bot$, and the value $\bot$ is only assigned on line 10, but only if $A(v) \notin \{0,1\}$.

Any $v \in V$ is assigned a value $x \in \{0,1\}$ on one of lines 15, 21, and 29. Consider the case of line 15. For this line to be reached, the condition on line 9 must evaluate to false, which means either $v = v_0$ or $A(v) \notin \{0,1\}$. If $A(v) \notin \{0,1\}$ then clearly the configuration v was not assigned before, so after assignment it has been assigned exactly once. If $v = v_0$, then the condition on line 13 is true, causing early termination and thus no further assignment of values to v. A similar argument holds for both line 21 and 29. □

Lemma 2. *Each edge $e = (v,T) \in E$ is picked on line 8 at most $\mathcal{O}(|V| + |E|)$ times.*

Proof. Let $e = (v,T) \in E$ be any edge. There are two ways for e to be added to W: either $e \in D(v')$ for some other configuration $v' \in T$ and was added on one of lines 15, 21, or 29, or v is the target of some other edge e' that is being explored and was added on line 31.

We consider each case in turn.

- If e is added at line 15, 21, or 29 then it is dependent on some $v' \in T$. By Lemma 1, this can happen at most once for each $v \in V$ and hence e can be added to W this way at most $|T|$ times.
- The edge e is added at line 31, i.e. there is a previously selected edge $e' = (v', T')$ such that $v \in T'$. Therefore, $A(v)$ was $\bot$ before this iteration of the algorithm, and is ? afterwards, so in order for e' to add e again in this way, line 10 must be reached while $A(v) = ?$. However, since $e' \in D(v)$, this can only happen if $A(v') \in \{0, 1\}$, in which case it cannot reach line 31. Hence e is added to W at line 31 at most once for each $e' = (v', T')$ such that $v \in T'$, or a total of $|E|$ times.

In conclusion, e is added to W at most $\mathcal{O}(|V| + |E|)$ times, and thus also selected from W $\mathcal{O}(|V| + |E|)$ times. □

As the **while**-loop contains no loops or recursion and DGs are finite, we get the following corollary.

Corollary 1. *Algorithm 1 terminates.*

We now state three important invariants of the main **while**-loop of the algorithm.

Lemma 3 (Loop invariant). *The following is a loop invariant for the* **while**-*loop of Algorithm 1.*

1. *For all $v \in V$, if $A(v) \in \{0, 1\}$ then $A(v) = A_{\min}(v)$.*
2. *For all $v \in V$ and $e = (u, T) \in E$, $e \in D(v)$ implies that $v \in T$.*
3. *For all $v \in V$, if $A(v) = ?$ then for all edges $e = (v, T) \in E$, either $e \in W$ or there exists $u \in V$ such that $e \in D(u)$ and $A(u) = ?$.*

Proof. We prove these invariants in turn.

1. Let $e = (v, T)$ be the edge picked on line 8. We focus on the places where values are assigned. If $A(v)$ is assigned on Line 15, then we immediately know that for all $u \in T$, $A(u) = 1$ so the algorithm sets $A(v) = 1$. By the invariant, we also know for all $u \in T$ that $A_{\min}(u) = 1$, so $A_{\min}(v) = 1$. If $A(v)$ is assigned on either of lines 21 or 29, then there are no successors to v that could backpropagate value 1, so $A_{\min}(v) = 0$ hence the invariant holds.
2. The only place where $D(u)$ is updated for any $u \in V$ is on line 24. On this line, $D(u)$ is updated to include edge $e = (v, T)$, but since u was selected on line 23, $u \in T$ so the invariant is maintained.
3. Let $v \in V$ be a configuration such that $A(v) = ?$, and assume that the edge selected on line 8 is $e = (v, T)$. If the algorithm assigns $A(v) = x$ where $x \in \{0, 1\}$ in this iteration then the invariant clearly holds, so we are left with the following cases.
   - If the condition on line 9 is true, then $A(v)$ is set to $\bot$, so the invariant is preserved wrt. v, and furthermore for each $e' = (u, T') \in D(v)$ we have $A(u) \in \{0, 1\}$, so the invariant also holds for each such u.

– Otherwise on line 23 we pick a configuration $u \in T$ such that $A(u) \notin \{0,1\}$ and add e to $D(u)$. If $A(v) = ?$ at this point, then either $\mathrm{sucs}(u) = \emptyset$ so $A(u)$ is assigned 0 and e is added to W, in which case the invariant holds, or $A(u)$ is set to ? and each $e' \in \mathrm{sucs}(u)$ are added to the waiting list, so the invariant holds. □

With the help of the previous lemmas we can now prove the correctness of Algorithm 1.

Theorem 1 (Correctness). *Algorithm 1 terminates, with return value* 1 *iff* $A_{\mathrm{min}}(v_0) = 1$.

Proof. Termination is provided by Corollary 1, and clearly if $A(v_0)$ was assigned $x \in \{0,1\}$ while running the algorithm, then $A(v) = A_{\mathrm{min}}(v_0)$ by Condition 1. of Lemma 3.

We now argue that if the algorithm terminates on line 32, then $A_{\mathrm{min}}(v_0) = 0$. First we note that at this point $A(v_0) = ?$, as $A(v_0)$ is initialised to ? before the **while**-loop, and if $A(v_0)$ was assigned value $x \in \{0,1\}$ then we would have terminated early. To demonstrate that $A_{\mathrm{min}}(v_0) = 0$ we construct a assignment B defined as

$$B(v) = \begin{cases} 0 & \text{if } A(v) \in \{0,?\} \\ 1 & \textbf{if otherwise.} \end{cases}$$

In words, this assignment promotes ? to 0. We wish to prove that B is a fixed-point assignment in a subgraph of G that contains v_0. The region of interest is given by a set Q, which we define as the least set such that

- $v_0 \in Q$, and
- if $v \in Q$ then for all $u \in V$ such that there exists an edge $e = (v,T) \in D(u)$ and $A(u) = ?$ we have $u \in Q$.

In words, Q is the set of configurations reachable from v_0 via only configurations with value ?. We show that B is a fixed-point assignment over all $v \in Q$, i.e. for all $v \in V$ we have

$$B(v) = \bigvee_{(v,T) \in E} \bigwedge_{u \in T} B(u) \ .$$

For the sake of contradiction, assume that there is some configuration $v \in Q$ such that $B(v) = 0$ but there exists a hyperedge $e = (v,T)$ such that $B(u) = 1$ for all $u \in T$. Due to Condition 1. of Lemma 3, it cannot be the case that $B(v) = 0$ because $A(v) = 0$, so $A(v) = ?$. Thus by Condition 3. of Lemma 3 we have either that $e \in W$ or there exists $u \in V$ such that $e \in D(u)$ and $A(u) = ?$, but since the algorithm terminated outside the **while**-loop, $W = \emptyset$, hence $e \in D(u)$ for some u where $A(u) = ?$. By Condition 2. we also have $u \in T$. But we have $B(u) = 0$ and $u \in T$, which contradicts our assumption that for all $u \in T$ we have $B(u) = 1$.

Because of this, B is a fixed-point assignment on the subgraph induced by $Q \cup \{u \in V \mid v \rightarrow u \text{ for some } v \in Q\}$. Since A_{min} is the least fixed-point assignment and $B(v_0) = 0$, $A_{\mathrm{min}}(v_0) = 0$, so we are done. □

5 Implementation and Experiments

We implement Algorithm 1 as an extension to the verification engine `verifypn` of the tool TAPAAL [11], a Petri net model checker written in C++. In doing so, the nondeterminism at lines 8 and 23 needs to be resolved.

DFS vs. BFS. In order to resolve the nondeterminism at line 8, we need to select an $e \in W$ to process, or in other words, we should select a suitable data structure for representing the set W. We implement W as two sets, W_f containing edges inserted during the forward exploration, and W_b containing edges inserted due to the backpropagation. Edges are added to W_f on lines 1 and 31, and to W_b on lines 15, 21, and 29. When selecting an edge on line 8, we select first from W_b if possible, and from W_f otherwise. During our experiments, we experience that the underlying structure of W_b makes little difference, so we keep it as a stack. For W_f we evaluate two choices:
- W_f is a stack, which we refer to as DFS.
- W_f is a queue, which we refer to as BFS.

Eager vs. Lazy. At line 23, we need to select an $u \in T$ that is not fully assigned. We evaluate the following two options for partially resolving this nondeterminism.
- We can prioritise configurations u such that $A(u) = ?$, i.e. configurations that were visited before but not fully assigned. We refer to this choice as lazy.
- We can prioritise configurations u such that $A(u) = \bot$, i.e. configurations that were not yet explored. We refer to this choice as eager.

In either case, if there is more than one configuration with the prioritised value, we pick an arbitrary one.

We shall now experimentally evaluate the new algorithm and the different implementation choices described above. We evaluate it against the CTL benchmark of the Model Checking Contest (MCC) 2021 [20] data set, which consists of 1181 Petri nets which are each associated with 16 CTL formulae for a total of 18 896 problem instances, as well as against benchmarks of Petri net games detailed in [4] and [12]. We name the different versions by using the naming introduced above. If our improvement starting at line 9 is enabled, we append an asterisk (*) to the configuration name, otherwise the improvement is not enabled. For example, DFS-lazy* denotes that W_f is a stack, configurations with $A(v) = ?$ are prioritised and it is our improved version of the algorithm. We do not compare against other tools, since TAPAAL significantly outperformed other competitors of the MCC'21 [20] and MCC'22 [21] even without the present improvement. A reproducibility package containing the data, raw results, binaries, and scripts used for running and analysing the experiments can be found at [17].

5.1 CTL Benchmark

For the CTL evaluation, we run the engine on each formula in the benchmark with a time limit of 5 min and memory limit of 15 GiB per formula, using AM

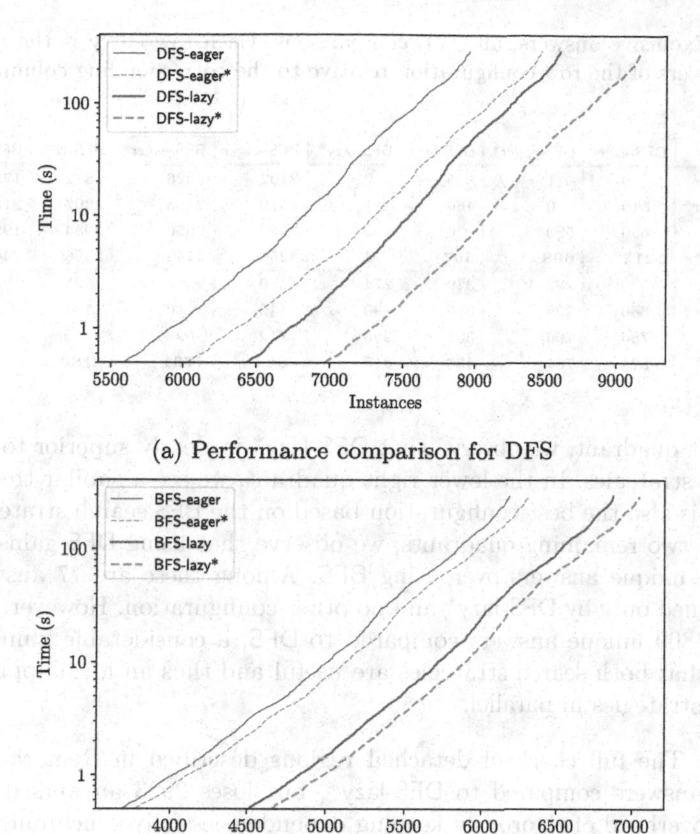

(a) Performance comparison for DFS

(b) Performance comparison for BFS

Fig. 4. Experiments for CTL model checking

Opteron 6376 Processors. Figures 4a and b show cactus plots, where for each
xperimental configuration, the runtime (on y-axis) for each instance is sorted
n ascending order independent of the other configurations, and the instances
re plotted in this order on the x-axis (we only show the most difficult instances
ere). Notice that the running times are plotted using a logarithmic scale. All
unning times refer only to time used in the verification algorithm. We observe
hat when using DFS, our new algorithm (shown in the plots as dashed lines)
nproves the average performance by about a factor 5 when using the lazy setup,
nd a factor 3 when using the eager setup. A more modest improvement can be
een also when using the BFS strategy and also here the eager strategy performs
etter. Overall, the DFS solves the largest number of CTL queries within the
min timeout.

Table 1 shows the number of unique answers between each pair of experimen-
l configurations, such that each row entry indicates the number of exclusive
swers compared to the configuration in the corresponding column. For exam-
e, DFS-lazy gains 590 answers relative to DFS-eager* but loses 460 answers. In

Table 1. Exclusive answers, all CTL comparisons. Each row entry is the number of unique answers of the row configuration relative to the corresponding column configuration.

	DFS-eager	DFS-eager*	DFS-lazy	DFS-lazy*	BFS-eager	BFS-eager*	BFS-lazy	BFS-lazy*
DFS-eager	0	1	5	0	2102	2026	1872	1788
DFS-eager*	516	0	460	1	2570	2460	2307	2183
DFS-lazy	650	590	0	0	2737	2656	2084	1995
DFS-lazy*	**1212**	**698**	**567**	**0**	3259	3148	2596	2441
BFS-eager	329	282	319	274	0	13	25	25
BFS-eager*	380	299	365	290	140	0	133	29
BFS-lazy	786	706	353	298	712	693	0	8
BFS-lazy*	874	754	436	315	**884**	**761**	**180**	**0**

the top left quadrant, we observe that DFS-lazy* is clearly superior to all other three DFS strategies. In the lower right quadrant, we see a similar trend where BFS-lazy* is also the best configuration based on the BFS search strategy. Comparing the two remaining quadrants, we observe that using DFS gains between 2000–3000 unique answers over using BFS. Among these are 77 answers that were obtained only by DFS-lazy* and no other configuration. However, BFS also gains 300–800 unique answers compared to DFS, a considerable number. This indicates that both search strategies are useful and thus an ideal approach will run these strategies in parallel.

Remark 3. The full check of detached regions described in Remark 2 has 16 exclusive answers compared to DFS-lazy*, but loses 2923 answers due to the excessive overhead of rigorously keeping dependencies fully synchronised.

5.2 Games Synthesis Benchmark

We further implemented our improved algorithm in the Petri net games verification engine, which is also a part of the `verifypn` engine. Based on the experience from CTL experiments, we only implemented the best performing configuration using the DFS search order. We evaluate this algorithm on the game synthesis benchmarks presented in [4] and [12].

The benchmark of [4] includes 6 scalable case studies. Each model in the case study is given 1 h and 32 GiB memory. Out of the 6 case studies, on 4 of them we observed minimal difference between the original and improved algorithm, while on the Producer/Consumer systems and on the model of the Lyngby Trai Station we noticed significant improvements up to two orders of magnitude, a shown in Fig. 5a and b.

The topology zoo benchmark of [12] (originally described in [19]) consist of 261 real network topologies of up to 700 nodes as well as nestings and con catenations of these, for a total of 1035 problem instances. Our evaluation i based on the problem of synthesis of network updates, encoded as Petri ne games. Figure 5c shows a cactus plot with the results. Our improved algorith gains about 80 answers over the certain-zero baseline algorithm and perform

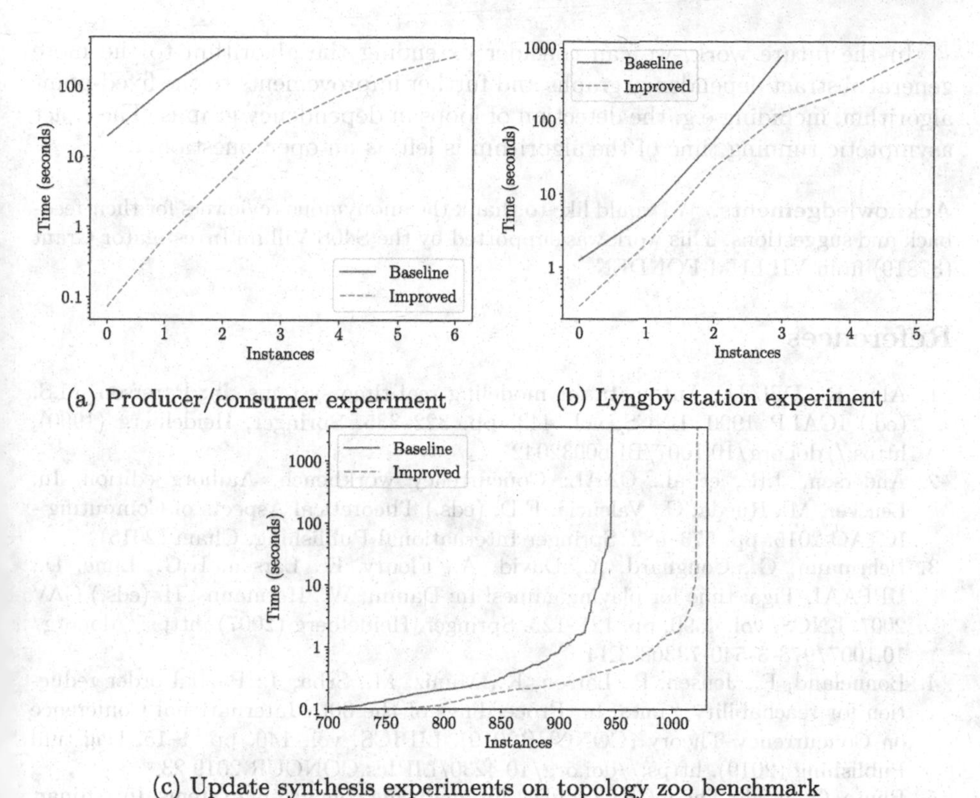

(a) Producer/consumer experiment (b) Lyngby station experiment

(c) Update synthesis experiments on topology zoo benchmark

Fig. 5. Petri games experiments

better by over an order of magnitude on harder problem instances. This demonstrates that our detached regions elimination technique is applicable to a range for dependency graphs coming from both the model checking domain as well as from strategy synthesis for games.

6 Conclusion

Ve presented a novel improvement to the local Liu and Smolka's algorithm used or verification of dependency graphs. Our algorithm detects detached regions 1 the dependency graph in order to speedup the performance of the fixed-point omputation. We proved that our improved algorithm is correct and provided its fficient implementation, as part of the tool TAPAAL, both for CTL model checkıg and Petri games synthesis. We evaluated the performance of our algorithm 1 benchmarks of Petri games and CTL formulae on Petri nets, demonstrating oticeable improvements in verification speed (compared to the state-of-the-art pproaches, including the most recent certain-zero algorithm) with only negligie overhead. We also observed that the depth-first search strategy is the most eneficial one, however, breadth-first search should be also considered as it can ovide a large number of unique answers.

In the future work, we can consider extending the algorithm to the more general abstract dependency graphs and further improvements to the fixed-point algorithm, including e.g. the detection of loops in dependency graphs. The exact asymptotic running time of the algorithm is left as an open question.

Acknowledgements. We would like to thank the anonymous reviewers for their feedback and suggestions. This work was supported by the S40S Villum Investigator Grant (37819) from VILLUM FONDEN.

References

1. Alur, R., Dill, D.: Automata for modeling real-time systems. In: Paterson, M.S. (ed.) ICALP 1990. LNCS, vol. 443, pp. 322–335. Springer, Heidelberg (1990). https://doi.org/10.1007/BFb0032042
2. Andersen, J.R., et al.: CAAL: Concurrency workbench, Aalborg edition. In: Leucker, M., Rueda, C., Valencia, F.D. (eds.) Theoretical Aspects of Computing - ICTAC 2015, pp. 573–582. Springer International Publishing, Cham (2015)
3. Behrmann, G., Cougnard, A., David, A., Fleury, E., Larsen, K.G., Lime, D.: UPPAAL-Tiga: time for playing games! In: Damm, W., Hermanns, H. (eds.) CAV 2007. LNCS, vol. 4590, pp. 121–125. Springer, Heidelberg (2007). https://doi.org/10.1007/978-3-540-73368-3_14
4. Bønneland, F., Jensen, P., Larsen, K., Muniz, M., Srba, J.: Partial order reduction for reachability games. In: Proceedings of the 30th International Conference on Concurrency Theory (CONCUR2019). LIPICS, vol. 140, pp. 1–15. Dagstuhl Publishing (2019). https://doi.org/10.4230/LIPIcs.CONCUR.2019.23
5. Bunte, O., et al.: The mCRL2 toolset for analysing concurrent systems. In: Vojnar, T., Zhang, L. (eds.) TACAS 2019. LNCS, vol. 11428, pp. 21–39. Springer, Cham (2019). https://doi.org/10.1007/978-3-030-17465-1_2
6. Cassez, F., David, A., Fleury, E., Larsen, K.G., Lime, D.: Efficient on-the-fly algorithms for the analysis of timed games. In: Abadi, M., de Alfaro, L. (eds.) CONCUR 2005. LNCS, vol. 3653, pp. 66–80. Springer, Heidelberg (2005). https://doi.org/10.1007/11539452_9
7. Clarke, E.M., Emerson, E.A.: Design and synthesis of synchronization skeletons using branching time temporal logic. In: Kozen, D. (ed.) Logic of Programs 1981. LNCS, vol. 131, pp. 52–71. Springer, Heidelberg (1982). https://doi.org/10.1007/BFb0025774
8. Dalsgaard, A.E., Enevoldsen, S., Larsen, K.G., Srba, J.: Distributed computation of fixed points on dependency graphs. In: Fränzle, M., Kapur, D., Zhan, N. (eds.) SETTA 2016. LNCS, vol. 9984, pp. 197–212. Springer, Cham (2016). https://doi.org/10.1007/978-3-319-47677-3_13
9. Dalsgaard, A.E., et al.: Extended dependency graphs and efficient distributed fixed-point computation. In: van der Aalst, W., Best, E. (eds.) PETRI NETS 2017 LNCS, vol. 10258, pp. 139–158. Springer, Cham (2017). https://doi.org/10.1007/978-3-319-57861-3_10
10. Dalsgaard, A.E., et al.: A distributed fixed-point algorithm for extended dependency graphs*. Fundamenta Informaticae **161**(4), 351–381 (2018)
11. David, A., Jacobsen, L., Jacobsen, M., Jørgensen, K.Y., Møller, M.H., Srba, J TAPAAL 2.0: integrated development environment for timed-arc petri nets. In Flanagan, C., König, B. (eds.) TACAS 2012. LNCS, vol. 7214, pp. 492–49 Springer, Heidelberg (2012). https://doi.org/10.1007/978-3-642-28756-5_36

12. Didriksen, M., et al.: Automatic synthesis of transiently correct network updates via petri games. In: Buchs, D., Carmona, J. (eds.) PETRI NETS 2021. LNCS, vol. 12734, pp. 118–137. Springer, Cham (2021). https://doi.org/10.1007/978-3-030-76983-3_7
13. Enevoldsen, S., Larsen, K.G., Mariegaard, A., Srba, J.: Dependency graphs with applications to verification. Int. J. Softw. Tools Technol. Transfer **22**(5), 635–654 (2020). https://doi.org/10.1007/s10009-020-00578-9
14. Enevoldsen, S., Guldstrand Larsen, K., Srba, J.: Abstract dependency graphs and their application to model checking. In: Vojnar, T., Zhang, L. (eds.) TACAS 2019. LNCS, vol. 11427, pp. 316–333. Springer, Cham (2019). https://doi.org/10.1007/978-3-030-17462-0_18
15. Enevoldsen, S., Larsen, K.G., Srba, J.: Extended abstract dependency graphs. Int. J. Softw. Tools Technol. Transfer **24**(1), 49–65 (2021). https://doi.org/10.1007/s10009-021-00638-8
16. Garavel, H., Lang, F., Mateescu, R., Serwe, W.: CADP 2011: a toolbox for the construction and analysis of distributed processes. Int. J. Softw. Tools Technol. Transfer **15**(2), 89–107 (2013)
17. Gjøl Jensen, P., Larsen, K.G., Srba, J., Jensen Ulrik, N.: Reproducibility package: elimination of detached regions in dependency graph verification (2023). https://doi.org/10.5281/zenodo.7712764
18. Jensen, J.F., Larsen, K.G., Srba, J., Oestergaard, L.K.: Efficient model-checking of weighted CTL with upper-bound constraints. Int. J. Softw. Tools Technol. Transfer **18**(4), 409–426 (2014). https://doi.org/10.1007/s10009-014-0359-5
19. Knight, S., Nguyen, H.X., Falkner, N., Bowden, R., Roughan, M.: The internet topology zoo. IEEE J. Sel. Areas Commun. **29**(9), 1765–1775 (2011). https://doi.org/10.1109/JSAC.2011.111002
20. Kordon, F., et al.: Complete Results for the 2020 Edition of the Model Checking Contest. http://mcc.lip6.fr/2021/results.php (2021)
21. Kordon, F., et al.: Complete Results for the 2022 Edition of the Model Checking Contest. http://mcc.lip6.fr/2022/results.php (2022)
22. Kozen, D.: Results on the propositional μ-calculus. In: Nielsen, M., Schmidt, E.M. (eds.) Automata, Languages and Programming. ICALP 1982. LNCS, vol. 140, pp. 348–359. Springer, Heidelberg (1982). https://doi.org/10.1007/bfb0012782
23. Liu, X., Ramakrishnan, C.R., Smolka, S.A.: Fully local and efficient evaluation of alternating fixed points. In: Steffen, B. (ed.) TACAS 1998. LNCS, vol. 1384, pp. 5–19. Springer, Heidelberg (1998). https://doi.org/10.1007/BFb0054161
24. Liu, X., Smolka, S.A.: Simple linear-time algorithms for minimal fixed points. In: Larsen, K.G., Skyum, S., Winskel, G. (eds.) Automata, Languages and Programming. ICALP 1998. LNCS, vol. 1443, pp. 53–66. Springer, Heidelberg (1998). https://doi.org/10.1007/bfb0055040
25. Mariegaard, A., Larsen, K.G.: Symbolic dependency graphs for $\text{PCTL}_{\leq}^{\geq}$ model-checking. In: Abate, A., Geeraerts, G. (eds.) FORMATS 2017. LNCS, vol. 10419, pp. 153–169. Springer, Cham (2017). https://doi.org/10.1007/978-3-319-65765-3_9
26. Milner, R.: Communication and concurrency. Prentice Hall International series in Computer Science, Prentice Hall (1989)
27. Tarski, A.: A lattice-theoretical fixpoint theorem and its applications. Pac. J. Math. **5**(2), 285–309 (1955)

Potency-Based Heuristic Search with Randomness for Explicit Model Checking

Emil G. Henriksen, Alan M. Khorsid, Esben Nielsen, Theodor Risager, Jiří Srba(✉), Adam M. Stück, and Andreas S. Sørensen

Department of Computer Science, Aalborg University, Aalborg, Denmark
srba@cs.aau.dk

Abstract. Efficient state-space exploration has a significant impact on reachability analysis in explicit model checking and existing tools use several variants of search heuristics and random walks in order to overcome the state-space explosion problem. We contribute with a novel approach based on a random search strategy, where actions are assigned dynamically (on-the-fly) updated potencies, changing according to the variations of a heuristic distance to the goal configuration as encountered during the state-space search. We implement our general idea into a Petri net model checker TAPAAL and document its efficiency on a large benchmark of Petri net models from the annual Model Checking Contest. The experiments show that our heuristic search outperforms the standard search approaches in multiple metrics and that it constitutes a worthy addition to the portfolio of classical search strategies.

1 Introduction

Heuristic search strategies are widely applied in areas such as pathfinding and planning [7], where the popular A* algorithm [6] is often used. Similarly in model checking, instead of naively exploring the state-space using, e.g., Breadth First Search (BFS) or Depth First Search (DFS) , it can be beneficial to use a heuristic search to navigate in the state-space. Heuristic search has also been succesfully applied in Petri net verification tools [5,11,12]. Since 2011, the annual Model Checking Contest (MCC) [9] has been held; here Petri net tools compete to solve a variety of problems, such as reachability analysis and deadlock detection. The tools ITS-Tools [11], SMPT [2] and LoLA [12] have implemented a random walk state-space exploration, and Tapaal [5] is using a random depth-first search. These tools have placed top three in the reachability category in MCC'21 [10] and MCC'22 [9], showing that randomness improves tool efficiency [12].

We propose a novel search heuristic, called *Random Potency-First Search* (RPFS), which combines a heuristic search based on distance function together with randomness in order to achieve a competitive-edge compared to the existing strategies. Our objective is to increase the likelihood of finding (not necessarily the shortest) trace to goal configurations. The unique property of RPFS is that while searching the state-space, it learns which transitions are more likely to contribute to achieving a given reachability goal and it dynamically modifies

G. Caltais and C. Schilling (Eds.): SPIN 2023, LNCS 13872, pp. 180–187, 2023.
https://doi.org/10.1007/978-3-031-32157-3_10

transition potencies (a number that expresses how likely a transition is to be selected during the search) according to the learned information. Such dynamic potency updates proved recently beneficial for guiding random Monte Carlo walks [1] but have not yet been explored for state-space search strategies.

We implement our RPFS search in the `verifypn` engine [8] of the tool TAPAAL [5], the best scoring tool in the reachability category at the most recent edition of MCC'22 [9]. We then benchmark RPFS against the existing search strategies present in TAPAAL on a large set of MCC'22 models. The results indicate a convincing performance of RPFS: it solves additional 512 queries compared to the second best search strategy, which is a significant 5% increase in the number of answered queries. We then rerun the TAPAAL competition script (using a portfolio of several different search strategies) used at MCC'22, by replacing the existing heuristic search with RPFS and obtain over 1% additional answers. As TAPAAL, the winner of MCC'22, solves close to 95% of all queries in the reachability benchmark, the additional 1% increase is very significant.

Preliminaries. Let $\mathbb{N}_0$ be the set of natural numbers including zero. A *Petri net* is a triple $N = (P, T, W)$ where P is a finite set of *places*, T is a finite set of *transitions* s.t. $T \cap P = \emptyset$, and $W : (P \times T) \cup (T \times P) \rightarrow \mathbb{N}_0$ is a *weighted flow function.* A *marking* on N is a function $M : P \rightarrow \mathbb{N}_0$ assigning a number of *tokens* to places. A transition $t \in T$ is *enabled* in M if $M(p) \geq W(p, t)$ for all $p \in P$. An enabled transition t in M can fire and produce a marking M', written $M \xrightarrow{t} M'$, where $M'(p) \stackrel{\text{def}}{=} M(p) - W(p, t) + W(t, p)$ for all $p \in P$.

Graphically, places are drawn as circles, transitions are drawn as rectangles, and arcs are drawn as arrows. Unless an arc is annotated by a number, its default weight is 1; we do not draw normal arcs with weight 0. Tokens are denoted as a number inside a place. A Petri net example is given in Fig. 1.

We are interested in the reachability of the standard cardinality queries φ where $\varphi ::= e \bowtie e \mid \varphi \wedge \varphi \mid \varphi \vee \varphi \mid \neg\varphi$ for $\bowtie \in \{\leq, <, =, \neq, >, \geq\}$ and where expressions e are given by $e ::= n \mid p \mid e + e \mid e - e \mid n \cdot e$ such that $n \in \mathbb{N}_0$ and $p \in P$.

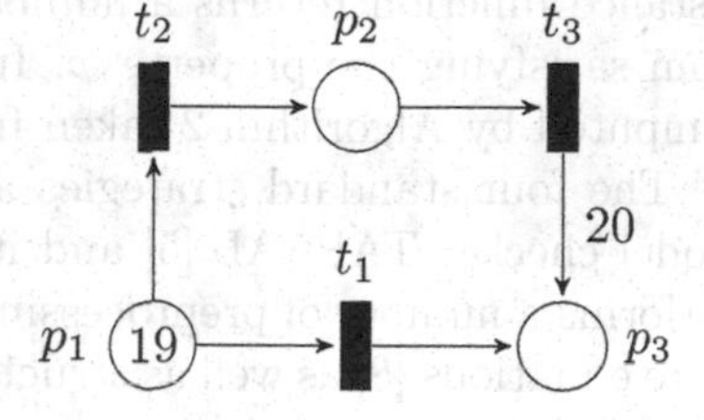

Fig. 1. Petri net example

An expression e in a marking M naturally evaluates to a number $eval(M, e)$ assuming that $eval(M, p) = M(p)$. The satisfaction relation $M \models \varphi$ is then defined in a straightforward way such that $M \models e_1 \bowtie e_2$ iff $eval(M, e_1) \bowtie eval(M, e_2)$.

For example in the net from Fig. 1, we can reach a marking satisfying the query $p_3 \geq 20$ by firing the transition t_2 followed by t_3.

Explicit State-Space Search

In order to solve the reachability problem, explicit model checkers perform a state-space search as depicted in Algorithm 1. The algorithm is generic in the

Algorithm 1: Generic Reachability Search Algorithm

```
Input: Petri net N, initial marking M0, proposition φ
Output: true if there is a reachable marking M s.t. M ⊨ φ, false otherwise
1  Function Generic-Reachability-Search(N, M0, φ):
2      if M0 ⊨ φ then return true ;
3      Initialise(Waiting, M0)            // Initialise Waiting with M0
4      Passed := {M0}
5      while Waiting is non-empty do
6          M := SelectAndRemove(Waiting)
           // Select and remove an element M from Waiting
7          for M' such that M -t-> M' where t is enabled in M do
8              if M' is not in Passed then
9                  if M' ⊨ φ then return true ;
10                 Passed := Passed ∪ {M'}
11                 Update(Waiting, t, M, M', φ)   // Update Waiting with M'
12     return false
```

way how it initialises the waiting set with M_0, updates the waiting set with the successor marking M', and selects and removes an element from the waiting set. In BFS strategy, the waiting set is implemented as a queue (FIFO) and in DFS it is implemented as a stack (LIFO). In RDFS, which stands for random DFS, the implementation is a stack where all successor markings M' of M are randomly shuffled before they are pushed to the waiting stack. A heuristic search strategy, called BestFS, implements the waiting set as a priority queue where markings that minimise the distance function $\texttt{Dist}(M, \varphi)$ are selected first. The distance function returns a number, estimating how far a given marking M is from satisfying the property φ. In the tool TAPAAL, the distance function is computed by Algorithm 2 taken from [8].

The four standard strategies are implemented for example in the Petri net model checker TAPAAL [5] and its engine `verifypn` [8]. The `verifypn` engine performs a number of preprocessing techniques like query simplification [4] using state equations [8] as well as structural reductions [3]. These techniques can solve a reachability query without executing the explicit state-space search. We evaluate the standard search strategies on Petri net models from MCC'22 [9]; each of the 1321 models are verified against 16 reachability cardinality queries, giving us the total of 21 136 problem instances. As 10 952 instances are solved without employing the explicit state-space search, we consider only the remaining 10 184 instances in our experiments on which we run all search strategies, enforcing 5 min timeout and a memory limit of 16 GB. The experiments are run on a CPU cluster using AMD EPYC 7551 32-core processors with clock speed at 2.5 GHz. A reproducibility package can be obtained at GitHub[1].

[1] https://github.com/theodor349/RPFS-reproducibility.

Algorithm 2: Distance Heuristics in TAPAAL, taken from [8]

Input: Marking M, cardinality query φ
Output: Nonnegative integer representing the distance of M from satisfying φ

1 **Function** Dist(M, φ):
2   **if** $\varphi = e_1 \bowtie e_2$ **then return** $\Delta(eval(M, e_1), \bowtie, eval(M, e_2))$;
3   **if** $\varphi = \varphi_1 \wedge \varphi_2$ **then return** Dist(M, φ_1) + Dist(M, φ_2) ;
4   **if** $\varphi = \varphi_1 \vee \varphi_2$ **then return** $min\{$Dist(M, φ_1), Dist$(M, \varphi_2)\}$;
5   **if** $\varphi = \neg(e_1 \bowtie e_2)$ **then return** $\Delta(eval(M, e_1), \overline{\bowtie}, eval(M, e_2))$;
6   **if** $\varphi = \neg(\varphi_1 \wedge \varphi_2)$ **then return** $min\{$Dist$(M, \neg\varphi_1)$, Dist$(M, \neg\varphi_2)\}$;
7   **if** $\varphi = \neg(\varphi_1 \vee \varphi_2)$ **then return** Dist$(M, \neg\varphi_1)$ + Dist$(M, \neg\varphi_2)$;
8   **if** $\varphi = \neg(\neg\varphi_1)$ **then return** Dist(M, φ_1) ;

where $\overline{\bowtie}$ is the dual arithmetical operation of $\bowtie$ (for example $\overline{>}$ is the notation for $\leq$) and where

$$\Delta(v_1, =, v_2) = |\ v_1 - v_2\ | \qquad \Delta(v_1, \neq, v_2) = \begin{cases} 1 & \text{if } v_1 = v_2 \\ 0 & \text{otherwise} \end{cases}$$

$$\Delta(v_1, <, v_2) = \max\{v_1 - v_2 + 1, 0\} \qquad \Delta(v_1, >, v_2) = \Delta(v_2, <, v_1)$$
$$\Delta(v_1, \leq, v_2) = \max\{v_1 - v_2, 0\} \qquad \Delta(v_1, \geq, v_2) = \Delta(v_2, \leq, v_1)$$

Table 1. Comparison of search strategies

Strategy	Total	Solved	Solved %	Fastest	Unique	Unique All
BFS	10184	7367	72.3 %	856	100	16
DFS	10184	8235	80.9 %	1691	14	7
RDFS	10184	8641	84.8 %	2452	218	41
BestFS	10184	8617	84.6 %	2358	177	31
Virtual Best	10184	9179	90.1 %	–	–	–
RPFS	10184	9153	89.9 %	2000	–	178
Virtual Best All	10184	9357	91.9 %	–	–	–

The upper part of Table 1 (ignore for now the rows with RPFS strategy and Virtual Best All, as well as the column Unique All) shows the comparison of the basic search strategies as implemented in TAPAAL's verification engine, the best tool at MCC'22 in the reachability category. The numbers of solved queries for each strategy indicate that the random DFS (RDFS) as well as the standard heuristic search (BestFS) are the most succesfull strategies, both solving almost 85% of all queries. These are also the two strategies that solve largest number of queries fastest. The number of unique answers (number of queries that a given strategy solved but none of the other three provided any answer) also indicates that the random and heuristic searches are the most beneficial ones. It is worth to note that BFS also obtains a significant number (100) of unique answers. This is due to the positive queries that have a relatively short witness trace,

```
Initialise(Waiting, M0) =
Waiting := {(M0, −)}
foreach transition t in T do
  └ Potencies(t) := 100
```

(a) Initialise

```
Update(Waiting, t, M, M', φ) =
Waiting := Waiting ∪ {(M', t)}
Potencies(t) := max{1, Potencies(t)+
        Dist(M, φ) − Dist(M', φ)}
```

(b) Update

```
SelectAndRemove(Waiting) =
n := 0
foreach transiton t in Waiting do
  | n := n + Potencies(t)
  | r := a random number in [0, 1]
  | if r ≤ Potencies(t)/n then
  |   └ t_best := t
(M, t_best) :=      arg min       Dist(M, φ)
               (M,t_best)∈Waiting
Waiting := Waiting \ (M, t_best)
return M
```

(c) Select and remove

Fig. 2. The pseudocode for RPFS

where BFS manages to find them but the other strategies miss these answers as they explore the state-space in a more depth-like manner.

These results indicate that one can consider to combine the heuristic search strategy with randomness, which we indeed tried by modifying the BestFS strategy so that we randomly select a marking from the priority queue among the first n markings that minimise the distance to the reachability query. The optimal value of n is around 100 where the performance peeks by solving 8761 instances (corresponding to 86.0% of all queries). In the next section, we shall present our idea of combining heuristic search with randomness by dynamically changing the probabilities of which transition to fire next during the state-space exploration. This novel method achieves even more significant performance improvement as it can solve almost 90% of all queries on its own.

3 Random Potency-First Search (RPFS)

We shall now describe our RPFS strategy where we assign potencies (positive numbers) to transitions. During the random search, transitions with higher potencies are more likely to be selected. We modify the transition potencies dynamically during the state-space search as we learn more information about which transitions decrease the distance to satisfying a given reachability query.

The RPFS algorithm is described in Fig. 2 where it instantiates the three primitives from Algorithm 1.

We initialise *Waiting* as a set of marking-transition pairs (M, t) representing a marking M that was reached by firing t (initially no transition is fired to reach M_0, so we use '-' here). The potencies of all transitions are set to 100. To select marking from *Waiting*, we first pick some transition t from the waiting set with the probability $\mathit{Potencies}(t)/\sum_{t' \text{ appearing in } \mathit{Waiting}} \mathit{Potencies}(t')$ and return marking that can be reached by firing t and minimises the distance function. For an efficient implementation of the selection of t, we use an algorithm described in [1]. Finally, the update function changes the potency of the transition t by the difference between the current distance and the new distance after firing t. Should the distance increase, the potency of t is lowered accordingly but we keep the potency of all transitions strictly positive. Implementation-wise, *Waiting* is implemented as a collection of priority queues, one for each transition. As such, for an element (M, t), the marking M is only stored in the priority queue for t. Because of this, we can select markings efficiently, as we do not need to search through all markings in the waiting set.

Let us consider the net from Fig. 1 where we ask about the reachability of the query $p_3 \geq 20$. The standard BestFS heuristics attempts to find a solution by repeatedly firing t_1, as this reduces the distance to satisfying the reachability query. First when all 19 tokens are in the place p_3, the search backtracks and explores the firing of transition t_2 that brings us to the goal (after firing also t_3). In total, 20 markings are explored before t_2 is fired. Because of the random choice in RPFS, the number of explored markings is not deterministic; we run RPFS 100 000 times on this example and noticed that it expanded on average 3.03 markings before it reached the goal. Hence the random choice allows us to explore with a certain probability also alternative search options, while still having a preference towards the more promising markings.

4 Experimental Evaluation of RPFS

We have rerun the RPFS strategy search in the identical experimental setup as presented in Sect. 2 and the results are summarised in Table 1. Our RPFS strategy solves 89.9% of all queries, which is almost the same as all four remaining strategies solve together (the virtual best of these strategies is 90.1%). In the number of unique answers over all five search strategies, RPFS got the largest number of 178 unique answers; in particular the number of unique answers for BFS dropped to only 16, which is a clear indication that RPFS is more efficient in finding short witness traces that the RDFS and BestFS have difficulties dealing with. The benchmark contains a large number of easy-to-solve instances, where RDFS and BestFS are faster due to a smaller overhead compared to running RPFS. Still, on 2000 instances RPFS provides the fastest answer. We also experimented with different ways of modifying the transition potencies (e.g. by altering them by a constant value); these changes did not make any significant difference on the performance of RPFS. In Fig. 3 we can see the four standard search strategies, RPFS and virtual best (over all five strategies) for our experimental evaluation. Independently for each search strategy, all instances (on x-axis) are ordered by their running times (on y-axis) and plotted in the comparison graphs. Note that y-axis uses the logarithmic scale. We can see that RPFS is clearly the best performing and it is closer to the virtual best (over all strategies) than to the BestFS and RDFS that show very similar performance.

In the final experiment, we run the competition script of TAPAAL on all models in the reachability category. The script uses a portfolio management, where several search strategies and other optimisations are run in parallel on

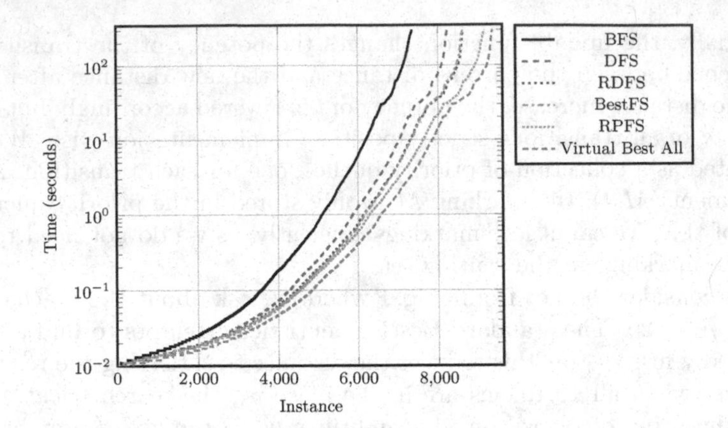

Fig. 3. A performance plot comparing standard search strategies and RPFS

Table 2. Solved instances from the total of 42272 (21136 for each subcategory)

	BestFS	RPFS	BestFS %	RPFS %
Cardinality	20308	20471	96.08%	96.85%
Fireability	19541	19894	92.45%	94.12%
Total	39849	40365	94.27%	95.49%

up to 4 cores for one hour on all 16 queries. In the script, we replace the use of BestFS with our new RPFS strategy and the number of answered queries can be seen in Table 2. We can observe a nontrivial increase in the answers, both in the cardinality queries and even more in the fireability ones (that are internally transformed into cardinality queries). In total, the RPFS heuristic provided an improvement of 1.22% over the performance of last year's winner of MCC'22.

5 Conclusion

We described RPFS, a novel idea of random, heuristic-based search strategy where transition potencies are dynamically updated during the state-space search. We instantiated this search strategy to the Petri net framework, however, we believe that the idea can be applicable to other formalisms as well. The RPFS strategy provides a significant performance boost to the existing strategies and it is now implemented in the state-of-the-art tool TAPAAL. In future work we shall study how to further improve the performance of RPFS e.g. by guiding the search according to the solutions to state-equations obtained by linear programming, or by turning it into a variant of A* search.

Acknowledgements. We thank the anonymous reviewers for their comments and suggestions and Peter G. Jensen for his help with the experimental setup.

References

1. Aagreen, E.F.L., Hansen, T.B.S., Herum, R.E.N., A.Jensen, F., Jensen, M.T.: Extending petri nets with transition weights to improve model-checking using Monte Carlo simulations (2022). https://doi.org/10.5281/zenodo.7690715, DAT8 Project report at Aalborg University, Denmark
2. Amat, N., Berthomieu, B., Dal Zilio, S.: On the combination of polyhedral abstraction and SMT-based model checking for petri nets. In: Buchs, D., Carmona, J. (eds.) PETRI NETS 2021. LNCS, vol. 12734, pp. 164–185. Springer, Cham (2021). https://doi.org/10.1007/978-3-030-76983-3_9
3. Bønneland, F., Dyhr, J., Jensen, P., Johannsen, M., Srba, J.: Stubborn versus structural reductions for Petri nets. J. Log. Algeb. Methods Program. **102**(1), 46–63 (2019)
4. Bønneland, F., Dyhr, J., Jensen, P.G., Johannsen, M., Srba, J.: Simplification of CTL formulae for efficient model checking of petri nets. In: Khomenko, V., Roux, O.H. (eds.) PETRI NETS 2018. LNCS, vol. 10877, pp. 143–163. Springer, Cham (2018). https://doi.org/10.1007/978-3-319-91268-4_8
5. David, A., Jacobsen, L., Jacobsen, M., Jørgensen, K.Y., Møller, M.H., Srba, J.: TAPAAL 2.0: integrated development environment for timed-arc petri nets. In: Flanagan, C., König, B. (eds.) TACAS 2012. LNCS, vol. 7214, pp. 492–497. Springer, Heidelberg (2012). https://doi.org/10.1007/978-3-642-28756-5_36
6. Hart, P.E., Nilsson, N.J., Raphael, B.: A formal basis for the heuristic determination of minimum cost paths. IEEE Trans. Syst. Sci. Cybern. **4**(2), 100–107 (1968)
7. Hoffmann, J., Nebel, B.: The FF planning system: fast plan generation through heuristic search. J. Artif. Intell. Res. **14**, 253–302 (2001)
8. Jensen, J.F., Nielsen, T., Oestergaard, L.K., Srba, J.: TAPAAL and reachability analysis of P/T nets. Trans. Petri Nets Other Model. Concurr. **11**, 307–318 (2016)
9. Kordon, F., et al.: Complete Results for the 2022 Edition of the Model Checking Contest (2022). http://mcc.lip6.fr/2022/results.php
10. Kordon, F., et al.: Complete Results for the 2021 Edition of the Model Checking Contest (2021). http://mcc.lip6.fr/2021/results.php
11. Thierry-Mieg, Y.: Symbolic model-checking using ITS-tools. In: Baier, C., Tinelli, C. (eds.) TACAS 2015. LNCS, vol. 9035, pp. 231–237. Springer, Heidelberg (2015). https://doi.org/10.1007/978-3-662-46681-0_20
12. Wolf, K.: Petri net model checking with LoLA 2. In: Khomenko, V., Roux, O.H. (eds.) PETRI NETS 2018. LNCS, vol. 10877, pp. 351–362. Springer, Cham (2018). https://doi.org/10.1007/978-3-319-91268-4_18

GPUexplore 3.0: GPU Accelerated State Space Exploration for Concurrent Systems with Data

Anton Wijs(✉) and Muhammad Osama

Eindhoven University of Technology, Eindhoven, The Netherlands
{a.j.wijs,o.m.m.muhammad}@tue.nl

Abstract. GPUexplore 3.0 is an explicit state space exploration tool that runs entirely on a graphics processing unit (GPU), and supports models of concurrent systems with data variables. We discuss its workflow and modelling language, present several design decisions regarding work distribution and retrieval, and experimentally evaluate the impact of those decisions. Our tool achieves acceleration up to 115× and 28× compared to single- and four-core LTSmin, respectively. It currently checks for deadlocks, with verification of temporal logic formulae planned for the near future.

Keywords: Explicit state space exploration · finite-state machines · GPU

1 Introduction

Graphics processing units (GPUs) are successfully applied for a plethora of applications, ranging from fluid dynamics [3] to deep learning [21], to drastically speed up computations, and in the last decade, also have contributed to accelerating explicit-state model checking [2,5,8,23,34,35,38–41,43], term rewriting [12], symbolic model checking [24,28], and SAT solving [25–27,29,44,45]. Initially, they were used to speed up specific aspects of model checking, such as probability computations for probabilistic model checking [4,17,36], successor generation [10,11], property checking after the state space had been constructed on the CPU [1], and counter-example construction [42]. GPUexplore [39,41] was the first tool to explicitly explore state spaces entirely on a GPU, without any computations performed by the CPU. Soon, other tools followed, most notably Grapple [8], a swarm-based explorer, ParaMoc, a model checker fo pushdown automata [35], and VoxLogicA-GPU [5], a spatial model checker t reason about (medical) images.

In GPUexplore 2.0, each individual process in a concurrent system i encoded as a Labelled Transition System (LTS) [20] that is stored in memor as a sparse matrix [32]. However, this does not allow efficient encodings of co current systems *with variables*. For example, consider a system with two 32-b integer variables x and y, and one process in which y is assigned the value of x

G. Caltais and C. Schilling (Eds.): SPIN 2023, LNCS 13872, pp. 188–197, 2023.
https://doi.org/10.1007/978-3-031-32157-3_11

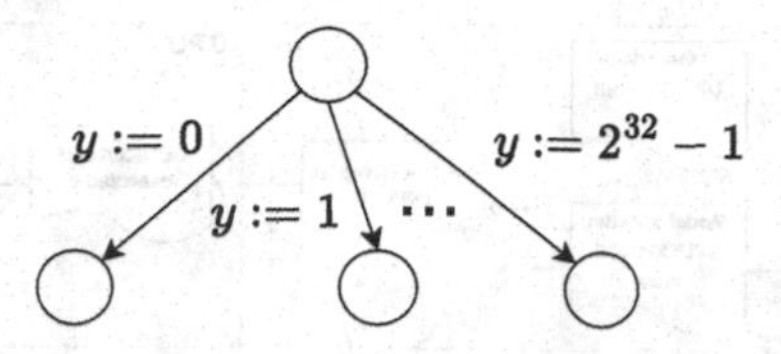

Fig. 1. Handling variables in GPUEXPLORE 2.0.

some point. Allowing for all possible values, GPUEXPLORE 2.0 requires that the LTS describing this process contains at least 2^{32} states, just to distinguish all possible values assigned to y (see Fig. 1). Thus, as variables are introduced, the matrices grow rapidly. Furthermore, GPU state space exploration tools are not user-friendly. Providing input is tedious, requiring manually setting up low-level descriptions of models [8,41] or using a chain of other tools [35,41].

For GPUEXPLORE 3.0, we wanted to change that, and directly support a richer modelling language. The tool altogether avoids storing the input model in memory. To make this possible and high-performant, we developed a code generator that produces GPU code specific for verifying a given input model. Conceptually, this is similar to how SPIN transforms PROMELA models to pan code [14]. GPUEXPLORE 3.0 is the first GPU tool to apply this. Although, at a high level, its exploration mechanism has remained the same, its code base has drastically changed, the result of three years of work. The tool can check for deadlocks, and we plan to add support for Linear Temporal Logic (LTL).

In fact, this code generation extends further than is typical for CPU-based model checkers such as SPIN. With the introduction of variables in input models, states grow in size. GPUEXPLORE 3.0 is the first GPU tool *in general*, to maintain a *tree database* [18,19]. The states of input models are stored as *binary trees*, which enables effective data sharing. This requires code generation of the storage functions, as the structure and size of trees depend on the input model, and tree storage has to be performed in a non-recursive way, since recursion is letrimental to GPU performance. In addition, it is the first GPU tool to apply *Cleary compression* [6,7] to store tree roots, allowing 64-bit roots to be stored in 32-bit integers. This combination means that once a few million states have been tored, the storage of each additional state requires only 32 bits, *independent* of ts size. This is completely novel for GPU hash tables in general [22].

In this paper, we present the workflow and modelling language of GPU-XPLORE 3.0, discuss design decisions regarding work distribution and work tching, and we experimentally evaluate the impact of those decisions.

Workflow and Modelling Language

orkflow. Figure 2 presents the workflow of GPUEXPLORE 3.0. The tool cepts models written in the *Simple Language of Communicating Objects* LCO) [31], described in more detail later. Given an input model, a code

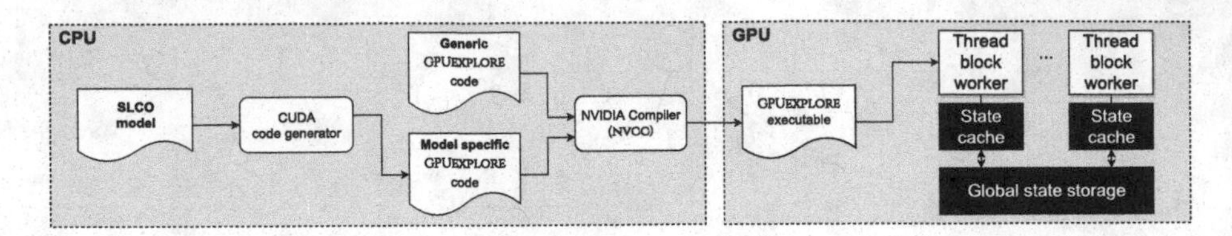

Fig. 2. The workflow of GPUEXPLORE 3.0.

generator, implemented in PYTHON using TEXTX [9] and JINJA2[1], produces *model-specific* code written in NVIDIA's CUDA C++. This code entails next-state computation functions, i.e., functions that given a system state s, produce the successor system states that can be reached from s by executing a transition. SLCO models consist of a finite number of Finite State Machines (FSMs) that concurrently execute transitions. In the model-specific code, one next-state computation function is produced for each FSM in the model, allowing for the successor states of a single state to be constructed in parallel, with the functions executed by different threads. This parallel construction of successors does not influence the correctness of the exploration: together, the threads end up exploring all possible execution paths of the input SLCO model. In addition, the model specific code involves the handling of state trees, the structure and size of which depend on the input model.

Combined with GPUEXPLORE's *generic* code, which implements the control flow and hash table, the code is compiled using NVIDIA's NVCC compiler. The resulting executable is suitable for CUDA-compatible GPUs with at least compute capability 7.0 and 24 GB global memory. GPUEXPLORE launches many thread blocks of 512 threads each. Each block uses fast on-chip memory[2] to maintain a *state cache*, in which the resulting successors of next state computation are stored, before the block checks the global tree database, access to which is much slower. The database is located in global memory, which is the largest on a GPU (24 GB in a Titan RTX). GPUEXPLORE operates in iterations. In each iteration, each block obtains states that require processing, computes successors, and stores them in the tree database if needed. This is repeated until all discovered states have been processed.

SLCO. SLCO models contain a finite number of FSMs and have global and FSM-local variables. The types Boolean, Integer, Byte, and arrays of those types are supported. Each FSM contains a finite number of transitions between its states, with one executable (atomic) statement associated with each transition. Statements can refer to all shared variables and those of the corresponding FSM, and are of the form $[e; x_0 := e_0; \ldots; x_n := e_n]$. The x_i's are references to variables or array elements and each e_i is an expression of the same type as x_i, and is constructed by combining references to variables and/or array elements using the typical logical and numerical operators, and e is a Boolean expression. Statemen

[1] https://palletsprojects.com/p/jinja/.
[2] On a Titan RTX, used for this work, on-chip memory is 49,152 bytes in size.

```
model M {
  classes
     GlobalClass {
       variables
       Byte c := 1
       Integer x1, x2
       state machines
       S0 {
         initial Q
         states  R S
         transitions
         Q -> R {
           [c < 20; x1 := c]
         }
         R -> S {
           [x1 := x1 + c]
         }
         ...
       }
       S1 {
         ...
       }
     }

   objects globalObject:
       GlobalClass()
}
```

(a) SLCO model M

```
switch (current_state) {
  case 0:
  {
    // Allocate register memory
    // to process transition(s).
    elem_inttype buf32_0, buf32_1;
    indextype bufaddr_0, bufaddr_1;

    // Q --{ [ c < 20; x1 := c ] }--> R

    mode = STORED;
    // Fetch values of unguarded variables.
    part1 = get_vectorpart(node_index, 0);
    part2 = get_vectorpart(node_index, 1);
    get_globalObject_c(&buf32_0, part1,
        part2);
    // Statement computation.
    if (buf32_0 < 20) {
       target = 1;
       buf32_1 = buf32_0;
       mode = (mode == STORED ? TO_CACHE :
           TO_GLOBAL);
       while (mode != STORED
              && mode != GLOBAL_STORED) {
       // Store new state vector in the
          cache
       // or the global hash table.
       ...
```

(b) Generated CUDA code for M

Fig. 3. Translating SLCO models into CUDA.

$[x_0 := e_0; \ldots; x_n := e_n]$ is shorthand for $[\mathbf{true}; x_0 := e_0; \ldots; x_n := e_n]$, and $[e]$ is a statement without assignments. The semantics of a transition is (informally) as follows: if e of its statement evaluates to **true**, the assignments $x_0 := e_0; \ldots x_n := e_n$ can be executed in sequence, by which the variables are updated, and the FSM atomically changes state, moving from the source state of the transition to the target state. If multiple transitions can be executed, the FSM changes state non-deterministically. Regarding concurrency, SLCO has an interleaving semantics.

Figure 3 presents an example SLCO FSM and part of the generated code. The FSM is taken from a translation of the `adding.1` model from the BEEM benchmark suite [30]. It has three process states, `Q` being the initial state. The transition statements refer to two of the three variables in the model, `c` and `x1`.

Given a system state and an FSM, a GPU thread generates successors by executing the corresponding next-state computation function. This function contains a big `switch` statement to consider the execution of transitions based on the current state of the FSM. In the example, if this FSM state, fetched from the system state and stored in the variable `current_state`, is `Q` (encoded as `0`), then the thread will retrieve the value of `c`, and store it in the variable `buf32_0`, located in thread-local *register memory*. If this value is smaller than 20, the target FSM state is set to `1` (`R`) and the register variable `buf32_1`, associated with

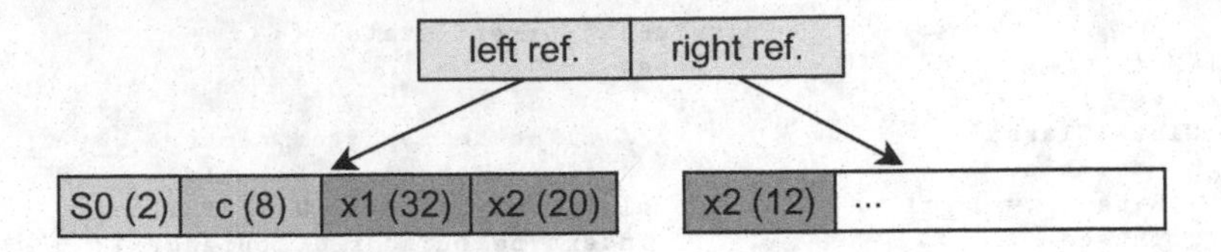

Fig. 4. State tree example.

x1, is assigned the value of buf32_0, i.e., c. Next, the thread will construct the new successor state by combining the original state with the new values, and store the new state in the state cache or, if it is full, the global tree database.

System states are stored as binary trees, with each tree node being a 64-bit integer. Each node can store up to 62 bits of information, with 2 bits used for bookkeeping. Figure 4 shows an example of such a tree, for the FSM given in Fig. 3. The leaf on the left stores the current state of FSM S0, which requires 2 bits, followed by the values of the variables. For x2, the value is stored in two leaves. The root consists of two references to the leaves, each requiring 29 bits to refer to a position in a hash table for non-roots with 2^{29} entries, but can be physically stored as a 32-bit integer in a separate root table with 2^{32} entries, using Cleary compression [6]. For this, invertible hash functions h_i are used. Given a node n, $h_i(n)$ provides both an address a and a remainder n' of less than 32 bits, which is stored at a. Given a remainder n' stored at a, n can be reconstructed by computing $h_i^{-1}(a, n')$. More details about the state storage can be found in [37].

3 Work Distribution and Retrieval Optimisations

Work Distribution over Thread Blocks. Each thread block has a *work tile* of a fixed size, which is filled with states that require processing at the start of each iteration. As the block produces new states, it can claim them for processing in the next iteration, but as soon as it produces more states than it can fit in its tile, the remaining work is left in the tree database for other blocks. In this way, GPUexplore does not apply *work stealing*, but rather *work sharing*.

Work Distribution Inside a Block. Inside a block, threads execute in groups of 32 threads, called *warps*. Each warp has a single program counter, hence the threads run in lock-step. This means that whenever the threads in a warp *diverge*, i.e., execute different lines of code, performance deteriorates, as the whole warp has to move over a line of code if at least one thread needs to execute it. For GPUexplore 3.0, we experimented with several options for work distribution in a block. At the top in Fig. 5, a strategy is visualised called *thread-to-FSM*. In this example, the model contains three FSMs, and their FSM states for the i-th state in the work tile are named S_0^i, S_1^i and S_2^i. The colours represent different warps. For ease of presentation, we assume that a warp has four threads. Given that

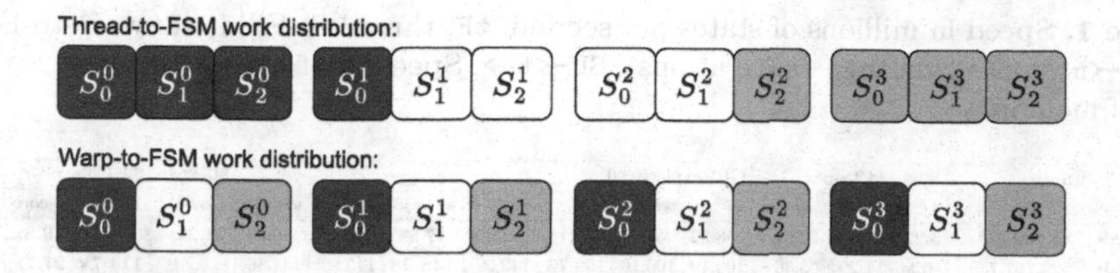

Fig. 5. Thread group tile processing strategies.

for each FSM, we have a separate next-state function, this distribution leads to the threads inside a warp diverging when they call the next-state function for their FSM. Another distribution is illustrated at the bottom of Fig. 5, called *warp-to-FSM*. Now, all threads in a warp are assigned the same FSM, resulting in those threads calling the same function using different data.

Reducing thread divergence can be taken further. Two threads that execute the same function but have different current FSM states still diverge, as they execute different `switch` cases (see Fig. 3). To minimise this, we sort the tile for each warp w.r.t. the current FSM state of its designated FSM. This results in all states with the same FSM state for the designated FSM being placed at consecutive positions in the tile, thereby stimulating that threads with consecutive IDs access states with the same current FSM state. Since the work tile is sufficiently small for the threads in a warp to store the tile in their combined register memory, sorting can be done in the register memory with *intra-warp bitonic merge sort* [15], using fast intra-warp instructions.

Multiple Iterations. Another optimisation is to execute *multiple iterations* in each exploration function call. GPUEXPLORE calls an exploration function to execute one or more next state iterations. Shared memory is wiped once a function execution finishes. With multiple iterations, a block can reuse the trees in its state cache constructed in one iteration, for exploration in the next one.

4 Tool Evaluation

Our code generator[3] can be launched with `python slcotogpuexplore.py <input-model>.slco [options]`. It takes an SLCO model as input and produces CUDA code. Several `options` can be given such as selecting a work distribution scheme or specifying the number of iterations per kernel launch. The code can be compiled with CUDA 11+ to produce an executable `gpuexplore` that can be launched with `./gpuexplore [-k <#ITERATIONS>]`.

For evaluation, we used SLCO models translated from a representative subset of the BEEM benchmark suite [30]. We scaled up some models, marked with '+'. For all experiments, we used CUDA 11.4, and a machine with a 4-CORE CPU i7-7700 operating at 3.6 GHz, 32GB RAM, and a Titan RTX GPU, running LINUX MINT 20.

[3] GPUEXPLORE is available for download here: https://bit.ly/3CUXTY8.

Table 1. Speed in millions of states per second. `tF`: thread-to-FSM, `wF`: warp-to-FSM, `wFs (+<n>)`: `wF + sorting (+ n` iterations), `SU-<to>`: Speedup of `wFs+30` vs. `<to>`, -O.M.-: out of memory.

Model	States	Spin	LTSmin		GPUexplore 3.0							SU-Spin	SU-LTSmin		SU-tF
		4-core	1-core	4-core	tF	wF	wFs	wFs+10	wFs+30	wFs+50	wFs+70	4-core	1-core	4-core	
adding.20+	84,709,120	3.22	1.40	3.94	58.02	55.65	57.18	**83.36**	77.89	67.12	59.60	**24.2x**	**55.8x**	**19.8x**	**1.2x**
adding.50+	529,767,730	-O.M.-	1.29	5.36	106.19	100.10	102.73	143.09	**148.28**	145.99	144.86	–	**114.7x**	**27.7x**	**1.4x**
anderson.6	18,206,917	1.36	0.67	1.31	9.58	13.80	16.02	31.58	31.57	**31.82**	31.71	**23.2x**	**47.2x**	**24.1x**	**3.3x**
anderson.7	538,699,029	-O.M.-	0.38	-O.M.-	7.93	15.43	20.95	**20.95**	19.78	19.75	19.68	–	**52.5x**	–	**2.5x**
at.5	31,999,440	1.50	0.61	1.88	14.05	23.79	28.73	36.74	36.54	**37.19**	36.58	**24.4x**	**60.4x**	**19.4x**	**2.6x**
at.6	160,589,600	0.87	0.66	2.39	14.39	27.76	38.34	**40.83**	40.56	40.59	40.62	**46.7x**	**61.9x**	**17x**	**2.8x**
at.7	819,243,816	-O.M.-	0.63	2.37	8.91	17.15	23.42	**23.60**	23.16	23.19	23.09	–	**36.7x**	**9.8x**	**2.6x**
bakery.5	7,866,401	2.57	0.62	0.90	7.52	7.71	7.46	11.29	19.02	**20.15**	19.98	**7.4x**	**30.9x**	**21x**	**2.7x**
bakery.7	29,047,471	2.59	0.76	1.62	8.47	9.10	9.06	20.80	29.12	30.98	**31.13**	**11.2x**	**38.5x**	**18x**	**3.7x**
bakery.8+	841,696,300	1.27	0.65	2.44	13.06	20.85	29.71	34.11	34.21	**34.31**	34.04	**27x**	**52.5x**	**14x**	**2.6x**
elevator2.3	7,667,712	1.10	0.46	0.99	3.48	3.32	3.24	5.98	6.06	**6.20**	6.10	**5.5x**	**13.1x**	**6.2x**	**1.8x**
elevator2.4+	91,226,112	0.56	0.57	1.95	2.97	3.74	**3.79**	3.22	3.28	3.33	3.34	**5.8x**	**5.8x**	**1.7x**	**1.1x**
elevator2.5+	1,016,070,144	-O.M.-	0.45	1.63	1.72	**1.88**	**1.88**	1.85	1.83	1.83	1.82	–	**4.1x**	**1.1x**	**1.1x**
frogs.4	17,443,219	2.23	0.50	1.42	7.37	10.06	9.75	11.13	**11.43**	11.32	11.26	**5.1x**	**22.9x**	**8x**	**1.5x**
frogs.5	182,772,126	1.05	0.70	2.63	6.45	9.63	9.61	10.27	**10.31**	10.23	10.18	**9.8x**	**14.6x**	**3.9x**	**1.6x**
lamport.6	8,717,688	1.38	0.49	1.10	5.07	5.20	5.09	17.94	27.35	**27.99**	27.80	**19.9x**	**55.6x**	**25x**	**5.5x**
lamport.7	38,717,846	1.82	0.62	1.98	11.00	18.13	23.04	33.50	34.47	34.45	**34.55**	**18.9x**	**55.5x**	**17.4x**	**3.1x**
lamport.8	62,669,317	1.78	0.80	2.19	10.73	18.55	25.45	34.31	34.92	35.12	**35.35**	**19.7x**	**43.9x**	**15.9x**	**3.3x**
loyd.3	239,500,800	-O.M.-	0.61	2.34	43.35	45.91	43.25	50.63	50.46	50.89	**51.04**	–	**82.3x**	**21.6x**	**1.2x**
mcs.5	60,556,519	0.62	0.42	1.49	12.07	19.44	24.26	29.98	**30.44**	30.34	30.25	**49.5x**	**72.1x**	**20.4x**	**2.5x**
peterson.5	131,064,750	1.62	0.73	2.44	11.75	21.13	28.44	**31.61**	31.28	30.76	30.70	**19.3x**	**43.1x**	**12.8x**	**2.6x**
peterson.6	174,495,861	0.76	0.68	2.45	12.05	21.04	30.47	**33.72**	33.58	33.31	33.19	**44.4x**	**49.4x**	**13.7x**	**2.8x**
peterson.7	142,471,098	1.50	0.72	2.27	10.17	20.93	22.37	**25.74**	25.44	25.21	25.21	**17x**	**35.4x**	**11.2x**	**2.5x**
phils.7	71,934,773	0.30	0.23	0.76	1.87	4.59	5.57	**5.66**	5.64	5.61	5.59	**19x**	**24.4x**	**7.4x**	**3x**
phils.8	43,046,720	0.36	0.28	0.79	2.64	9.07	8.96	**9.35**	9.27	9.21	9.17	**25.7x**	**33.5x**	**11.8x**	**3.5x**
szymanski.5	79,518,740	1.57	0.50	1.82	7.07	12.15	17.02	**19.03**	18.34	18.31	18.35	**11.7x**	**37x**	**10.1x**	**2.6x**
Average		1.43	0.63	2.00	15.30	19.85	22.99	29.63	**30.55**	30.20	29.81	**20.7x**	**44x**	**14x**	**3x**

Table 1 shows the results, comparing the impact of the presented options with four-core SPIN 6.5.1 [13] and single- and four-core LTSMIN 3.0.2 [16]. We only enabled state compression and basic reachability (without property checking) in those tools, to favour fast exploration of large state spaces. As GPUEXPLORE 3.0 does not yet have support for on-the-fly reduction methods, such as partial-order reduction [23], these have been disabled for all tools. Since LTSMIN scales near-linearly with the number of cores [33], the results indicate how many cores LTSMIN needs to be as fast as GPUEXPLORE. The best speeds are highlighted in bold. Overall, warp-to-FSM with sorting and 30 iterations is most successful.

Table 2. Millions of states per second for GPUEXPLORE 3.0 vs. version 2.0.

Tool	anderson.6	anderson.7	lamport.8	peterson.5	peterson.6	peterson.7	szymanski.5
2.0	15.863	-O.M.-	33.063	16.874	16.705	13.581	**26.454**
3.0	**34.111**	**22.326**	**35.387**	**32.331**	**34.902**	**26.183**	18.357

Finally, in Table 2, we compared GPUEXPLORE 3.0 with version 2.0 on th Titan RTX. In the comparison, we used all BEEM models for which correspondin GPUEXPLORE 2.0 models exist: `anderson.6` and `.7`, `lamport.8`, `peterson.`

.6 and .7 and `szymanski.5`. GPUEXPLORE 2.0 ran out of memory on the `anderson.7` model while GPUEXPLORE 3.0 was able to explore all models with an average acceleration of 1.8×. A comparison with GRAPPLE is discussed in a recent paper [37].

References

1. Barnat, J., Bauch, P., Brim, L., Češka, M.: Designing fast LTL model checking algorithms for many-core GPUs. JPDC **72**(9), 1083–1097 (2012). https://doi.org/10.1016/j.jpdc.2011.10.015
2. Bartocci, E., DeFrancisco, R., Smolka, S.A.: Towards a gpgpu-parallel spin model checker. In: SPIN 2014, pp. 87–96. ACM, New York (2014). https://doi.org/10.1145/2632362.2632379
3. Bertolli, C., Betts, A., Mudalige, G., Giles, M., Kelly, P.: Design and performance of the OP2 library for unstructured mesh applications. In: Alexander, M., et al. (eds.) Euro-Par 2011. LNCS, vol. 7155, pp. 191–200. Springer, Heidelberg (2012). https://doi.org/10.1007/978-3-642-29737-3_22
4. Bošnački, D., Edelkamp, S., Sulewski, D., Wijs, A.: Parallel probabilistic model checking on general purpose graphics processors. STTT **13**(1), 21–35 (2011). https://doi.org/10.1007/s10009-010-0176-4
5. Bussi, L., Ciancia, V., Gadducci, F.: Towards a spatial model checker on GPU. In: Peters, K., Willemse, T.A.C. (eds.) FORTE 2021. LNCS, vol. 12719, pp. 188–196. Springer, Cham (2021). https://doi.org/10.1007/978-3-030-78089-0_12
6. Cleary, J.: Compact hash tables using bidirectional linear probing. IEEE Trans. Comput. **c-33**(9), 828–834 (1984). https://doi.org/10.1109/TC.1984.1676499
7. Darragh, J., Cleary, J., Witten, I.: Bonsai: a compact representation of trees. Softw. Pract. Exp. **23**(3), 277–291 (1993). https://doi.org/10.1002/spe.4380230305
8. DeFrancisco, R., Cho, S., Ferdman, M., Smolka, S.A.: Swarm model checking on the GPU. Int. J. Softw. Tools Technol. Transf. **22**(5), 583–599 (2020). https://doi.org/10.1007/s10009-020-00576-x
9. Dejanović, I., Vaderna, R., Milosavljević, G., Vuković, Ž: TextX: a python tool for domain-specific language implementation. Knowl.-Based Syst. **115**, 1–4 (2017). https://doi.org/10.1016/j.knosys.2016.10.023
10. Edelkamp, S., Sulewski, D.: Efficient explicit-state model checking on general purpose graphics processors. In: van de Pol, J., Weber, M. (eds.) SPIN 2010. LNCS, vol. 6349, pp. 106–123. Springer, Heidelberg (2010). https://doi.org/10.1007/978-3-642-16164-3_8
11. Edelkamp, S., Sulewski, D.: External memory breadth-first search with delayed duplicate detection on the GPU. In: van der Meyden, R., Smaus, J.-G. (eds.) MoChArt 2010. LNCS (LNAI), vol. 6572, pp. 12–31. Springer, Heidelberg (2011). https://doi.org/10.1007/978-3-642-20674-0_2
12. van Eerd, J., Groote, J.F., Hijma, P., Martens, J., Osama, M., Wijs, A.: Innermost many-sorted term rewriting on gpus. Sci. Comput. Program. **225**, 102910 (2023). https://doi.org/10.1016/j.scico.2022.102910
13. Holzmann, G.J.: Parallelizing the spin model checker. In: Donaldson, A., Parker, D. (eds.) SPIN 2012. LNCS, vol. 7385, pp. 155–171. Springer, Heidelberg (2012). https://doi.org/10.1007/978-3-642-31759-0_12
14. Holzmann, G.: The model checker spin. IEEE Trans. Softw. Eng. **23**(5), 279–295 (1997). https://doi.org/10.1109/32.588521

15. Hou, K., Liu, W., Wang, H., Feng, W.: Fast segmented sort on gpus. In: Gropp, W.D., Beckman, P., Li, Z., Cazorla, F.J. (eds.) ICS, pp. 12:1–12:10. ACM (2017). https://doi.org/10.1145/3079079.3079105
16. Kant, G., Laarman, A., Meijer, J., van de Pol, J., Blom, S., van Dijk, T.: LTSmin: high-performance language-independent model checking. In: Baier, C., Tinelli, C. (eds.) TACAS 2015. LNCS, vol. 9035, pp. 692–707. Springer, Heidelberg (2015). https://doi.org/10.1007/978-3-662-46681-0_61
17. Khan, M.H., Hassan, O., Khan, S.: Accelerating SpMV multiplication in probabilistic model checkers using GPUs. In: Cerone, A., Ölveczky, P.C. (eds.) ICTAC 2021. LNCS, vol. 12819, pp. 86–104. Springer, Cham (2021). https://doi.org/10.1007/978-3-030-85315-0_6
18. Laarman, A.: Optimal compression of combinatorial state spaces. Innov. Syst. Softw. Eng. **15**, 235–251 (2019). https://doi.org/10.1007/s11334-019-00341-7
19. Laarman, A., van de Pol, J., Weber, M.: Parallel recursive state compression for free. In: Groce, A., Musuvathi, M. (eds.) SPIN 2011. LNCS, vol. 6823, pp. 38–56. Springer, Heidelberg (2011). https://doi.org/10.1007/978-3-642-22306-8_4
20. Lang, F.: Refined interfaces for compositional verification. In: Najm, E., Pradat-Peyre, J.-F., Donzeau-Gouge, V.V. (eds.) FORTE 2006. LNCS, vol. 4229, pp. 159–174. Springer, Heidelberg (2006). https://doi.org/10.1007/11888116_13
21. Le, Q.V., Ngiam, J., Coates, A., Lahiri, A., Prochnow, B., Ng, A.Y.: On optimization methods for deep learning. In: Getoor, L., Scheffer, T. (eds.) ICML, pp. 265–272. Omnipress (2011)
22. Lessley, B.: Data-parallel hashing techniques for GPU architectures. IEEE Trans. Parallel Distrib. Syst. **31**(1), 237–250 (2019). https://doi.org/10.1109/TPDS.2019.2929768
23. Neele, T., Wijs, A., Bošnački, D., van de Pol, J.: Partial-order reduction for GPU model checking. In: Artho, C., Legay, A., Peled, D. (eds.) ATVA 2016. LNCS, vol. 9938, pp. 357–374. Springer, Cham (2016). https://doi.org/10.1007/978-3-319-46520-3_23
24. Osama, M.: GPU Enabled Automated Reasoning. Ph.D. thesis, Eindhoven University of Technology (2022). ISBN: 978-90-386-5445-4
25. Osama, M., Gaber, L., Hussein, A.I., Mahmoud, H.: An efficient SAT-based test generation algorithm with GPU accelerator. J. Electron. Test. **34**(5), 511–527 (2018). https://doi.org/10.1007/s10836-018-5747-4
26. Osama, M., Wijs, A.: Parallel SAT simplification on GPU architectures. In: Vojnar, T., Zhang, L. (eds.) TACAS 2019. LNCS, vol. 11427, pp. 21–40. Springer, Cham (2019). https://doi.org/10.1007/978-3-030-17462-0_2
27. Osama, M., Wijs, A.: SIGmA: GPU accelerated simplification of SAT formulas. In Ahrendt, W., Tapia Tarifa, S.L. (eds.) IFM 2019. LNCS, vol. 11918, pp. 514–522 Springer, Cham (2019). https://doi.org/10.1007/978-3-030-34968-4_29
28. Osama, M., Wijs, A.: GPU acceleration of bounded model checking wit ParaFROST. In: Silva, A., Leino, K.R.M. (eds.) CAV 2021. LNCS, vol. 12760, pp 447–460. Springer, Cham (2021). https://doi.org/10.1007/978-3-030-81688-9_21
29. Osama, M., Wijs, A., Biere, A.: SAT solving with GPU accelerated inprocessing In: TACAS 2021. LNCS, vol. 12651, pp. 133–151. Springer, Cham (2021). https:/ doi.org/10.1007/978-3-030-72016-2_8
30. Pelánek, R.: BEEM: benchmarks for explicit model checkers. In: Bošnački, D Edelkamp, S. (eds.) SPIN 2007. LNCS, vol. 4595, pp. 263–267. Springer, Heidelbe (2007). https://doi.org/10.1007/978-3-540-73370-6_17

31. de Putter, S., Wijs, A., Zhang, D.: The SLCO framework for verified, model-driven construction of component software. In: Bae, K., Ölveczky, P.C. (eds.) FACS 2018. LNCS, vol. 11222, pp. 288–296. Springer, Cham (2018). https://doi.org/10.1007/978-3-030-02146-7_15
32. Saad, Y.: Iterative Methods for Sparse Linear Systems. SIAM, Philadelphia (2003)
33. van der Vegt, S., Laarman, A.: A parallel compact hash table. In: Kotásek, Z., Bouda, J., Černá, I., Sekanina, L., Vojnar, T., Antoš, D. (eds.) MEMICS 2011. LNCS, vol. 7119, pp. 191–204. Springer, Heidelberg (2012). https://doi.org/10.1007/978-3-642-25929-6_18
34. Wei, H., Chen, X., Ye, X., Fu, N., Huang, Y., Shi, J.: Parallel model checking on pushdown systems. In: ISPA/IUCC/BDCloud/SocialCom/SustainCom, pp. 88–95. IEEE (2018). https://doi.org/10.1109/BDCloud.2018.00026
35. Wei, H., Ye, X., Shi, J., Huang, Y.: ParaMoC: A Parallel Model Checker for Pushdown Systems. In: ICA3PP. LNCS, vol. 11945, pp. 305–312. Springer (2019). https://doi.org/10.1007/978-3-030-38961-1_26
36. Wijs, A.J., Bošnački, D.: Improving GPU sparse matrix-vector multiplication for probabilistic model checking. In: Donaldson, A., Parker, D. (eds.) SPIN 2012. LNCS, vol. 7385, pp. 98–116. Springer, Heidelberg (2012). https://doi.org/10.1007/978-3-642-31759-0_9
37. Wijs, A., Osama, M.: A GPU tree database for many-core explicit state space exploration. In: Sankaranarayanan, S., Sharygina, N. (eds.) TACAS 2023. LNCS, vol. 13993, pp. 684–703. Springer, Cham (2023). https://doi.org/10.1007/978-3-031-30823-9_35
38. Wijs, A.: BFS-based model checking of linear-time properties with an application on GPUs. In: Chaudhuri, S., Farzan, A. (eds.) CAV 2016. LNCS, vol. 9780, pp. 472–493. Springer, Cham (2016). https://doi.org/10.1007/978-3-319-41540-6_26
39. Wijs, A., Bošnački, D.: GPUexplore: many-core on-the-fly state space exploration using GPUs. In: Ábrahám, E., Havelund, K. (eds.) TACAS 2014. LNCS, vol. 8413, pp. 233–247. Springer, Heidelberg (2014). https://doi.org/10.1007/978-3-642-54862-8_16
40. Wijs, A., Bošnački, D.: Many-core on-the-fly model checking of safety properties using GPUs. Int. J. Softw. Tools Technol. Transf. **18**(2), 169–185 (2015). https://doi.org/10.1007/s10009-015-0379-9
41. Wijs, A., Neele, T., Bošnački, D.: GPUexplore 2.0: unleashing GPU explicit-state model checking. In: Fitzgerald, J., Heitmeyer, C., Gnesi, S., Philippou, A. (eds.) FM 2016. LNCS, vol. 9995, pp. 694–701. Springer, Cham (2016). https://doi.org/10.1007/978-3-319-48989-6_42
42. Wu, Z., Liu, Y., Liang, Y., Sun, J.: GPU accelerated counterexample generation in LTL model checking. In: Merz, S., Pang, J. (eds.) ICFEM 2014. LNCS, vol. 8829, pp. 413–429. Springer, Cham (2014). https://doi.org/10.1007/978-3-319-11737-9_27
43. Wu, Z., Liu, Y., Sun, J., Shi, J., Qin, S.: GPU accelerated on-the-fly reachability checking. In: ICECCS 2015, pp. 100–109 (2015). https://doi.org/10.1109/ICECCS.2015.21
44. Youness, H., Osama, M., Hussein, A., Moness, M., Hassan, A.M.: An effective SAT solver utilizing ACO based on heterogenous systems. IEEE Access **8**, 102920–102934 (2020). https://doi.org/10.1109/ACCESS.2020.2999382
45. Youness, H.A., Ibraheim, A., Moness, M., Osama, M.: An efficient implementation of ant colony optimization on gpu for the satisfiability problem. In: PDP, pp. 230–235. IEEE (2015). https://doi.org/10.1109/PDP.2015.59

Author Index

Caltais and C. Schilling (Eds.): SPIN 2023, LNCS 13872, p. 199, 2023.
s://doi.org/10.1007/978-3-031-32157-3

Printed in the United States
by Baker & Taylor Publisher Services

Printed in the United States
by Baker & Taylor Publisher Services